1ˢᵗ Edition

Basic College Math

LaTishia L. Jordan
Nashville Learning Center

ISBN 13: 978-1515111672
ISBN 10: 1515111679

Table of Contents

Chapter 1: Whole Numbers --- 4
    1.1 Whole Numbers --- 6
    1.2 Adding Whole Numbers --- 7
    1.3 Subtracting Whole Numbers -------------------------------------- 8
    1.4 Multiplying Whole Numbers ------------------------------------- 10
    1.5 Dividing Whole Numbers -- 12
    1.6 Order of Operations -- 14
    1.7 Prime Numbers and Prime Factorization -------------------------- 15
    Chapter 1 Review

Chapter 2: Fractions
    2.1a Least Common Multiple --------------------------------------- 21
    2.1b Least Common Denominator ----------------------------------- 21
    2.1c Greatest Common Factor ------------------------------------- 22
    2.2 Introduction to Fractions ------------------------------------ 23
    2.3 Equivalent Fractions -- 26
    2.4 Reducing Fractions -- 27
    2.5 Multiplying Fractions --------------------------------------- 27
    2.6 Dividing Fractions -- 28
    2.7 Adding Fractions -- 29
    2.8 Subtracting Fractions -------------------------------------- 30
    2.9 Adding, Subtracting, Multiplying, Dividing Improper and Mixed Numbers ------------- 31
    Chapter 2 Review --- 32

Chapter 3: Decimals
    3.1 Introduction to Decimals ----------------------------------- 35
    3.2 Adding Decimals -- 36
    3.3 Subtracting Decimals --------------------------------------- 37
    3.4 Multiplication of Decimals--------------------------------- 38
    3.5 Division of Decimals --------------------------------------- 38
    Chapter 3 Review -- 40

Chapter 4: Ratios and Proportions
    4.1 Rations --- 43
    4.2Rates --- 44
    4.3 Proportions -- 45
    Chapter 4 Review -- 46

Chapter 5: Percents
    5.1 Percents, Fractions, Decimals --------------------------- 49
    5.2 Rate, Base, Part -------------------------------------- 55
    5.3 Percent Problems: Proportion Method------------------- 57
    Chapter 5 Review --- 59

Cumulative Review: Chapters 1-5---62

Chapter 6: Geometry
    6.1 Types of Angles---67
    6.2 Triangles --72
    6.3 Pythagorean Theorem --75
    6.4 Quadrilaterals---77
    Chapter 6 Review --79

Chapter 7: Statistics
    7.1 Mean--82
    7.2 Median--83
    7.3 Mode ---84
    7.4 Range ---85
    Chapter 7 Review --86

Chapter 8: Measurement
    8.1 Converting Units ---88
    8.2 Area ---89
    8.3 Perimeter--96
    Chapter 8 Review -- 100

Chapter 9: Integers
    9.1 Integers --- 104
    9.2 Comparing and Ordering Integers --- 106
    9.3 Adding Integers-- 108
    9.4 Subtracting Integers-- 109
    9.5 Multiplying and Dividing Integers--- 110
    Chapter 9 Review --- 111

Chapter 10: An Introduction to Algebra
    10.1 Real Numbers -- 114
    10.2 Adding Real Numbers -- 117
    10.3 Subtracting Real Numbers -- 118
    10.4 Multiplying and Dividing Real Numbers -------------------------------------- 119
    10.5 Exponents and Order of Operations-- 120
    10.6 Algebraic Expressions--- 123
    10.7 Simplifying Algebraic Expressions Using Properties of Real Numbers------------- 124
    Chapter 10 Review-- 126

Chapter 11: Equation and Inequalities
    11.1 The Addition Property of Equality --- 129
    11.2 The Subtraction Property of Equality--- 130
    11.3 The Multiplication Property of Equality-------------------------------------- 131

11.4 The Division Property of Equality-- 132

11.5 Linear Equations --- 133

11.6 Formulas-- 134

11.7 Problem Solving-- 136

11.8 Solving Inequalities --- 138

Chapter 11 Review--- 139

Cumulative Review Chapters 6-11 -- 141

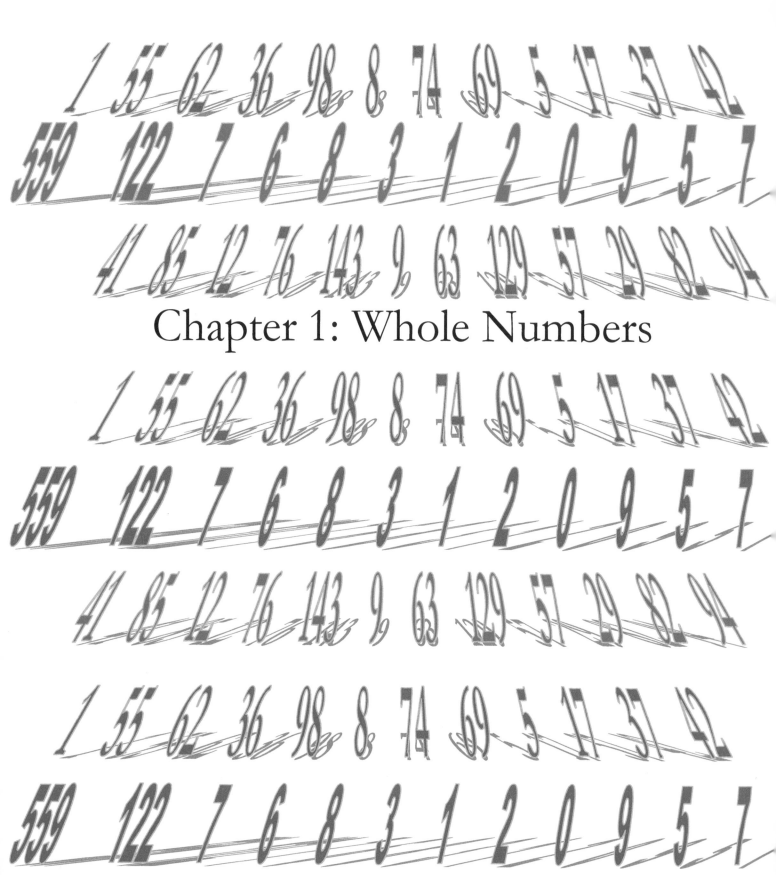

Chapter 1: Whole Numbers

1.1 Whole Numbers

Whole numbers also called natural numbers are the numbers used for counting including ordinal numbers.

The **whole numbers** are 0, 1, 2, 3, 4, 5, 6, 7, 8, 9, 10, 11, 12, and so on. They are used to answer questions such as:

- How many?
- How fast?
- How far?

Examples: 6, 15, 128, 1st, 5th, 139th

Whole numbers include 0 but do not include negative numbers, decimals, or fractions.

Guided Practice

Determine whether each number is a whole number

1) 3 _____

2) -14 _____

3) 0 _____

4) $\frac{1}{2}$ _____

5) 16th _____

6) .792 _____

7) -1.9 _____

8) 634 _____

9) $\frac{3}{8}$ _____

10) 91st _____

1.2 Adding Whole Numbers

First let's discuss the parts of an addition problem:

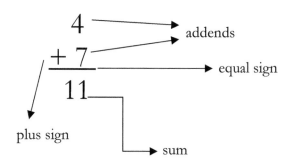

plus sign

When adding numbers with more than one digit add according to place value and carry extra digits to the next place value when needed. So start by adding your ones places.

1
123
+ 286
409

> Step 1: Add the ones place (3 + 6 = 9)
>
> Step 2: add the tens place (2 + 8 = 10); Place the 0 in the tens place and carry the 1 to the hundreds place.
>
> Step 3: Add the hundreds place (1 + 1 + 2 = 4)
>
> **Your sum or total is 409**

Practice 1.2

1) 265
    + 19

2) 367
    + 43

3) 784
    + 562

4) 626
    + 91

5) 624
    + 186

6) 483
    + 61

7) 483
    + 32

8) 806
    + 58

9) 526
    + 24

```
10)   274          12)   139          14)   301
    + 329              + 621              + 191


11)   124          13)   429          15)   903
    + 19               + 36               + 512
```

1.3 Subtracting Whole Numbers

First let's discuss the parts of a subtraction problem:

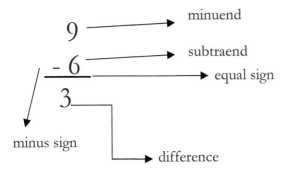

Subtracting with one digit is typically fairly easy because you are not borrowing. Things start to get complicated when you have multiple digits and you have to borrow. Let's start with the basic subtraction problem with no borrowing.

```
  25        subtract 5 − 2 = 3
- 12        subtract 2 − 1 = 1
  13
```

Subtraction with borrowing:

```
  3  16
   4̸6̸
-  29
   17
```

Step 1: Subtract 9 from 6 (6 − 9 is not possible so you must borrow from the tens place (the 4) and add the 10 the you borrow to the 6 to get 16 so that you can subtract 16 − 9 = 7.

Step 2: The 4 decreases by 1 and becomes a 3 and now you can subtract 3 − 2 = 1

Your difference is 17

Practice 1.3

1) 17
    − 6

2) 98
    − 17

3) 11
    − 8

4) 126
    − 58

5) 221
    − 74

6) 364
    − 132

7) 497
    − 248

8) 821
    − 396

9) 36
    − 19

10) 68
    − 29

11) 745
    − 123

12) 124
    − 63

13) 247
    − 193

14) 674
    − 526

15) 368
    − 59

1.4 Multiplying Whole Numbers

Multiplication is repeated addition, and it is written using a multiplication sign ×, which is read as "times."

We can write multiplication problems in horizontal or vertical form.

So let's talk about the parts of a multiplication problem:

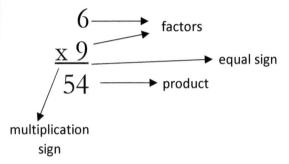

6 ⟶ factors
x 9 ⟶ equal sign
54 ⟶ product
multiplication sign

A raised dot • and parentheses ( ) are also used to write multiplication in horizontal form.

To multiply smaller numbers we simply use our multiplication facts but when we multiple larger whole numbers, we use a vertical form by stacking them with their corresponding place values lined up.

Then we make continual use of basic multiplication facts.

Smaller Numbers

Just use your facts!

```
   9
 x 3
  27
```

Now let's look at some larger numbers:

```
     1
    23
  x 14
    92
 + 230
   322
```

Step 1: Multiply the numbers in the ones place (3 x 4 = 12); place the 2 in the ones place in your product and your one in the tens place above the 2.

Step 2: Multiply the number in the tens place from your top number by the number in the ones place from your bottom number (2 x 4 = 8) and add that to the 1 from Step 1 (8 + 1 = 9).

Step 3: Move down one line in your answer and place a zero in your answer as a place holder.

Step 4: Multiply the 3 in the ones place in the top number by the 1 in the tens place in the bottom number (3 x 1 = 3).

Step 5: Multiply the 2 in the tens place in the top number by the 1 in the tens place in the bottom number (2 x 1 = 2).

Step 5: Add your answers and you get your product.

Your Product is 322

No matter how large your numbers are the method to solve them through multiplication does not change.

To multiply whole umbers that end in 0, you multiply your non-zero numbers and add your zeroes to the end.

$$\begin{array}{r} 80 \\ \times\ 5 \\ \hline 400 \end{array}$$
8 x 5 = 40 then tack the one 0 in the problem to the end of the 40 = 400

Practice 1.4

1) $\begin{array}{r} 17 \\ \times\ 6 \\ \hline \end{array}$

6) $\begin{array}{r} 364 \\ \times\ 132 \\ \hline \end{array}$

11) $\begin{array}{r} 745 \\ \times\ 123 \\ \hline \end{array}$

2) $\begin{array}{r} 98 \\ \times\ 17 \\ \hline \end{array}$

7) $\begin{array}{r} 497 \\ \times\ 248 \\ \hline \end{array}$

12) $\begin{array}{r} 124 \\ \times\ 63 \\ \hline \end{array}$

3) $\begin{array}{r} 11 \\ \times\ 8 \\ \hline \end{array}$

8) $\begin{array}{r} 821 \\ \times\ 396 \\ \hline \end{array}$

13) $\begin{array}{r} 247 \\ \times 193 \\ \hline \end{array}$

4) $\begin{array}{r} 126 \\ \times\ 58 \\ \hline \end{array}$

9) $\begin{array}{r} 36 \\ \times 19 \\ \hline \end{array}$

14) $\begin{array}{r} 674 \\ \times 526 \\ \hline \end{array}$

5) $\begin{array}{r} 221 \\ \times\ 74 \\ \hline \end{array}$

10) $\begin{array}{r} 68 \\ \times\ 29 \\ \hline \end{array}$

15) $\begin{array}{r} 368 \\ \times\ 59 \\ \hline \end{array}$

16) $\begin{array}{r} 70 \\ \times\ 14 \\ \hline \end{array}$

17) $\begin{array}{r} 100 \\ \times 25 \\ \hline \end{array}$

18) $\begin{array}{r} 30 \\ \times\ 9 \\ \hline \end{array}$

1.5 Dividing Whole Numbers

When you divide whole numbers, you are separating a quantity into equal-sized groups.

For example let's say you go out to dinner with a group of four friends and you decide you want to split the bill which is $60 equally. You would divide 60 into 5 equal groups to see how much each of you would pay.

You can write a division problem in many different ways such as using a division sign $60 \div 5$, a long division symbol $5\overline{)60}$ or a fraction bar $\frac{60}{5}$, $60/5$.

Now let's look at the parts of the division problem for each of these methods of writing the division problem.

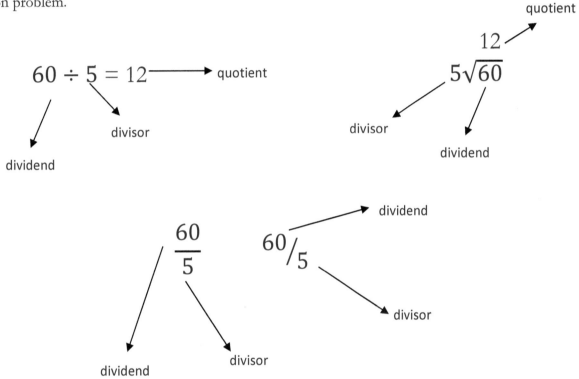

There are three special cases to consider when dividing.

1) When dividing by 1:
    When dividing something by 1, the answer is the original number. So, if the divisor is 1 then the quotient equals the dividend.

    Examples:
    $50 \div 1 = 50$
    $265 \div 1 = 265$

2) When dividing by 0:
   You cannot divide a number by 0. The answer to this question is undefined.

3) When your dividend equals divisor: If the dividend and the divisor are the same number (and not 0), then the answer is always 1.

   Examples:
   $60 \div 60 = 1$
   $32 \div 32 = 1$

If the quotient in a division problem is not a whole number, what is remaining is called the remainder.

Example:

$$
\begin{array}{r}
6 \\
3\overline{)20} \\
-18 \\
\hline
2
\end{array}
$$

2 is left over so that is your remainder!

Practice 1.5

1) $9\overline{)19}$

2) $87\overline{)261}$

3) $25\overline{)150}$

4) $4\overline{)11}$

5) $5\overline{)10}$

6) $74\overline{)592}$

7) $33\overline{)165}$

8) $3\overline{)8}$

9) $29\overline{)58}$

10) $60\overline{)540}$

11) $9\overline{)81}$

12) $98\overline{)784}$

13) $33\overline{)165}$

14) $73\overline{)292}$

15) $2\overline{)11}$

1.6 Order of Operations

The order of operation is the order in which you solve a problem that has multiple operations. The way that most of us remember the order of operations is by remembering the sentence:
Please **E**xcuse **M**y **D**ear **A**unt **S**ally. or PEMDAS

PEMDAS tells us that we solve problems with multiple operations in the following way:
Parentheses
Exponents
Multiplication
Division
Addition
Subtraction

Note: Multiplication and Division are interchangeable from left to right and Addition and Subtraction are interchangeable from left to right. This means that when you get to the point in which all of your parentheses (brackets) as well as your exponents have been solved then you can solve multiplication and division in the order in which they appear from left to right and then addition and subtraction in the order in which they appear from left to right.

Examples:

$4 + [-1(-2 - 1)]^2$
$4 + [-1\,(-3)]^2$
$4 + 9$
13

Step 1: Solve the parentheses and brackets first (solving from the inner most set out). (-2-1 = -3)

Step 2: Go to the brackets now $[-1\,(-3)]^2 = -1(-3) = 3^2 = 9$

Step 3: 4 + 9 = 13

$7 + (6 \times 5^2 + 3)$
$7 + (6 \times 25 + 3)$
$7 + 153$
160

Start with the exponents (because they are in the parentheses)
$5^2 = 25$
Multiply 6 × 25 = 150 + 3 = 153
Add 7 + 153 = 160

Practice 1.6

1) $3+10-2-(4\times2-3)$

2) $(1+(6\div3)^3)\times(3-2)$

3) $2\times(9+6-3-(5-2))$

4) $4\times5\div((8-5)\times1-2)$

5) $(8-5)^{(3\div3)}$

6) $7+6\div1-(4-2^2)$

7) $3\div3-1\,7\div(10\times1)$

8) $(6\times2+2)\div(2\times7-7)$

9) $8\div2+2^4-7+3$

10) $(2-(1\times6-(9-5)))\div4$

13) $8\times(5-4)-2\div(8\div8)$

11) $6-2+6\div2\times2\div3$

14) $(8-8)\times3\div(1+1)^4$

12) $7\div(6\div1\div6)\div(10\div10)$

15) $9\div(2\div(7\div(6+1))\div2)$

1.7 Prime Numbers and Prime Factorization

A prime number is a whole number greater than 1 that can be divided evenly only by one and itself. Below is a list of prime numbers up to 1000:

2	3	5	7	11	13	17	19	23	29
31	37	41	43	47	53	59	61	67	71
73	79	83	89	97	101	103	107	109	113
127	131	137	139	149	151	157	163	167	173
179	181	191	193	197	199	211	223	227	229
233	239	241	251	257	263	269	271	277	281
283	293	307	311	313	317	331	337	347	349
353	359	367	373	379	383	389	397	401	409
419	421	431	433	439	443	449	457	461	463
467	479	487	491	499	503	509	521	523	541
547	557	563	569	571	577	587	593	599	601
607	613	617	619	631	641	643	647	653	659
661	673	677	683	691	701	709	719	727	733
739	743	751	757	761	769	773	787	797	809
811	821	823	827	829	839	853	857	859	863
877	881	883	887	907	911	919	929	937	941
947	953	967	971	977	983	991	997		

Table 1.1

If you remember from Section 1.4 factors are numbers that you multiply together to get another number.

Prime factorization is finding which prime numbers multiply together to make the original number.

Example: What are the prime factors of 12?

$12 \div 2 = 6$ 2 is prime so we keep that as one of our prime factors.

6 is not prime so we can divide again $\qquad$ $6 \div 2 = 3$

The prime factors for $12 = 2 \times 2 \times 3$

This can also be written using exponents as $12 = 2^2 \times 3$

Example 2: What is the prime factorization of 19?

What are the factors of 19? 19, 1 so 19 is already prime (see Table 1.1).

We can also find prime factors using a factor tree. In a factor tree you solve the problem the same way by writing out the factors however each factor becomes a branch from the tree.

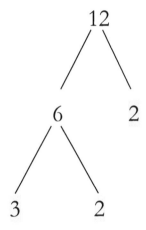

Example 3: Find the prime factors of 56

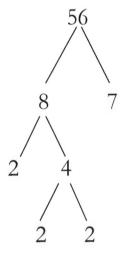

Prime Factorization of $56 = 2^3 \times 7$ or $2 \times 2 \times 2 \times 7$

Practice 1.7

1) 39

2) 86

3) 81

4) 46

5) 28

6) 55

7) 93

8)   194

9)   129

10) 180

11) 166

12) 44

13) 124

14) 112

15) 122

16) 140

17) 172

18) 135

19) 98

20) 196

Chapter 1 Review

Add

1) 81 + 78

2) 82 + 23

3) 73 + 36

4) 32 + 43

5) 94 + 34

6) 40 + 312

7) 239 + 90

8) 181 + 92

9) 196 + 25

10) 386 + 36

Subtract

11) 47 − 34

12) 94 − 40

13) 192 − 84

14) 77 − 30

15) 177 − 75

16) 97 − 48

17) 195 − 33

18) 266 − 140

19) 84 − 39

20) 71 − 59

Multiply

21) 16 x 69

22) 35 x 74

23) 12 x 47

24) 99 x 88

25) 25 x 10

26) 77 x 82

27) 70 x 47

28) 16 x 88

29) 82 x 51

30) 35 x 12

Divide

31) $90\sqrt{181}$

32) $36\sqrt{228}$

33) $74\sqrt{592}$

34) $12\sqrt{150}$

35) $19\sqrt{191}$

36) $24\sqrt{169}$

37) $48\sqrt{274}$

38) $16\sqrt{183}$

39) $21\sqrt{177}$

40) $28\sqrt{169}$

Find the Prime Factorization for the numbers below

41) 27

42) 196

43) 92

44) 160

45) 45

46) 158

47) 88

48) 40

49) 100

50) 56

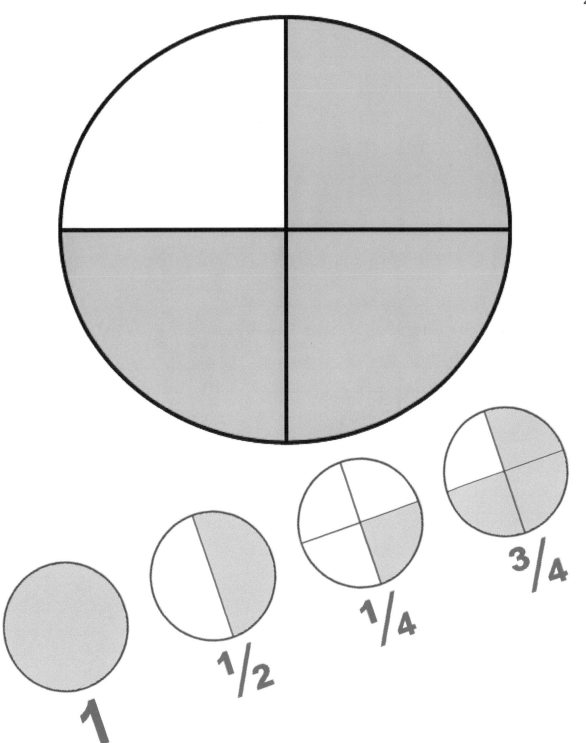

Chapter 2: Fractions

2.1a: Least Common Multiple (LCM):

Multiples are what we get after multiplying a number by another number. The Least Common Multiple is the smallest number that is a multiple of two or more numbers.

For example let's find the Least Common Multiple of 6 and 8

Multiples of 6 are: 6,12,18,24,30,36,42,48,54,60…..

Multiples of 8 are: 8, 16, 24, 32, 40, 48, 56, 64…..

Although 6 and 8 have more than one Common Multiple; the smallest number that 6 and 8 have in common is 24 so the Least Common Multiple of 6 and 8 is 24.

In order to find the least common multiple you just list the multiples of the numbers until you get your first match.

Example: Find the least common multiple of 4 and 5:

Multiples of 4: 4, 8, 12, 16, 20, 24, 28, 32 …..

Multiples of 5: 5, 10, 15, 20, 25, 30, 35, 40…..

As you can see the least common multiple of 4 and 5 is 20.

You can also find the least common multiple of more than two numbers. Let's find the least common multiple of 4, 12, and 6.

Multiples of 4: 4, 8, 12, 16, 20, 24, 28, 32 …..

Multiples of 6 are: 6,12,18,24,30,36,42,48,54,60…..

Multiples of 12: 12, 24, 36, 48, 60, 72, 84…..

As you can see the least common multiple of 4, 6, and 12 is 12.

2.1b: Least Common Denominator

In a fraction the denominator is your bottom number $\frac{numerator}{denominator}$

The denominator shows how many equal parts the item is divided into.

In order to find the least common denominator all we do is find the least common multiple.

Example: Find the least common denominator for:

$$\frac{1}{5}, \frac{2}{9}$$

Multiples of 5: 5, 10, 15, 20, 25, 30, 35, 40, 45, 50, 55, 60…

Multiples of 9: 9, 18, 27, 36, 45, 54, 63, 72……

The least common denominator is the same as the least common multiple which is 45.

We use the least common denominator when adding and subtracting fractions which is something you will visit later on in the chapter.

2.1c Greatest Common Factor (GCF):

The greatest common factor is the highest number that divides evenly into two or more numbers.

Factors are numbers that you multiply together to get another number. A number can have many factors.

> Example: The Greatest Common Factor for 10 and 15
>
> Factors of 10: 1, 2, 5, 10
>
> Factors of 15: 1, 3, 5, 15
>
> So the greatest common factor of 10, 15 is 5

The greatest common factor is used to simplify or reduce a fraction.

Practice 2.1:

Find the LCM of each of the following:

1) 2 and 7

2) 4 and 10

3) 4 and 5

4) 6 and 10

5) 4 and 12

6) 6 and 18

7) 12 and 3

8) 9 and 6

9) 10 and 3

10) 3 and 15

Find the GCF of each of the following:

1) 15 , 24

2) 40 , 15

3) 20 , 2

4) 10 , 30

5) 24 , 4

6) 10 , 15

7) 40 , 3

8) 2 , 10

Find the LCD of each of the following:

1) $\frac{6}{7}$ and $\frac{5}{13}$

2) $\frac{2}{3}$ and $\frac{7}{12}$

3) $\frac{1}{4}$ and $\frac{1}{2}$

4) $\frac{3}{8}$ and $\frac{2}{3}$

5) $\frac{4}{11}$ and $\frac{5}{8}$

6) $\frac{7}{10}$ and $\frac{1}{2}$

7) $\frac{6}{13}$ and $\frac{3}{5}$

8) $\frac{4}{9}$ and $\frac{1}{3}$

9) $\frac{7}{8}$ and $\frac{3}{4}$

10) $\frac{6}{11}$ and $\frac{4}{8}$

2.2 Introduction to Fractions

A fraction is a number that is used for measuring and written in the format below:

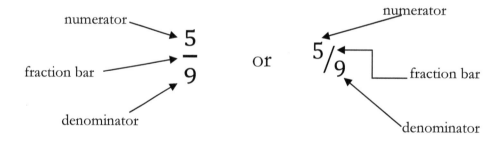

In a fraction you have a numerator, denominator, and a fraction bar which simply means divide. Your **common fraction** has a numerator and denominator that are both natural numbers.

The **denominator** is the number of equal parts into which the number 1 has been divided. So in the example above $\frac{5}{9}$ the number 1 is being divided into 9 parts. The **numerator** is the number of those equal parts that you are counting. Again using the above example we are counting 5 of the 9 equal parts.

A **proper fraction** is a fraction that is less than 1 that has a numerator smaller than the denominator.

An **improper fraction** is a fraction that is more than 1 and the numerator is larger than the denominator. For example $\frac{32}{9}$ is an improper fraction.

A **mixed number (fraction)** has a whole number and a proper fraction combined. For example $3\frac{5}{9}$ is a mixed fraction.

When the numerator is equal to the denominator as in $\frac{9}{9}$ the fraction is equal to 1.

To change an improper fraction to a mixed number or a whole number we divide the numerator by the denominator. The quotient becomes the whole number and the remainder becomes your new numerator with the denominator staying the same.

For example:

$$\frac{48}{9} = 5\frac{3}{9}$$

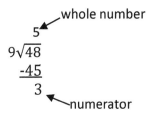

Practice 2.2a: Improper to Mixed

Change each improper fraction to a mixed fraction.

1) $\frac{30}{8}$

2) $\frac{42}{10}$

3) $\frac{13}{2}$

4) $\frac{39}{9}$

5) $\frac{22}{3}$

6) $\frac{40}{9}$

7) $\frac{22}{8}$

8) $\frac{41}{6}$

9) $\frac{17}{8}$

10) $\frac{57}{8}$

We can also work our way from mixed to improper by multiplying the denominator by the whole number and adding the numerator to get a new numerator and the denominator stays the same.

$$7\frac{2}{3} = \frac{23}{3}$$

Take the denominator 3 and multiply it by the whole number 7. (3 x 7 = 21)
Take your product 21 and add it to 2. (21 + 2 = 23)
23 is the new numerator and the denominator stays the same.

Practice 2.2b Mixed to Improper

Change each mixed fraction to an improper fraction.

1) $3\frac{1}{2}$

6) $7\frac{8}{9}$

2) $4\frac{5}{7}$

7) $3\frac{2}{5}$

3) $9\frac{2}{3}$

8) $2\frac{1}{7}$

4) $5\frac{5}{6}$

9) $5\frac{3}{4}$

5) $3\frac{9}{10}$

10) $2\frac{5}{9}$

2.3 Equivalent Fractions

Equivalent fractions are two or more fractions that have the same value but may not look the same. For example:

$$\frac{2}{3}, \frac{6}{9}, and \frac{12}{18}$$

are all equivalent fractions so they all equal each other.

$$\frac{2}{3} = \frac{6}{9} = \frac{12}{18}$$

You can find equivalent fractions by multiplying the numerator and denominator by the same number or dividing the numerator and denominator by the same number.

$$\frac{2}{3} \times \frac{3}{3} = \frac{6}{9} \times \frac{2}{2} = \frac{12}{18}$$

Practice 2.3

Write an equivalent fraction for each fraction.

1) $\frac{6}{7}$

3) $\frac{5}{6}$

5) $\frac{1}{9}$

7) $\frac{3}{10}$

2) $\frac{4}{5}$

4) $\frac{2}{3}$

6) $\frac{5}{9}$

8) $\frac{3}{4}$

2.4 Reducing Fractions

To reduce a fraction, divide the numerator and denominator by the largest number that can divide into both numbers exactly.

For instance let's say we have $\frac{3}{9}$ in order to reduce it to lowest terms we need to divide by the largest number that can be divided into both 3 and 9. The largest number that can be divided into both is 3.

$$\frac{3}{9} \div \frac{3}{3} = \frac{1}{3}$$

Practice 2.4

1) $\frac{16}{48}$

2) $\frac{14}{28}$

3) $\frac{7}{21}$

4) $\frac{12}{72}$

5) $\frac{18}{72}$

6) $\frac{15}{30}$

7) $\frac{8}{10}$

8) $\frac{13}{39}$

9) $\frac{30}{40}$

10) $\frac{17}{51}$

2.5 Multiplying Fractions

Multiplying fractions is one of the easiest operations to solve. In order to multiply fractions you multiply the numerators, multiply the denominators, and reduce the fraction.

$$\frac{3}{4} \times \frac{5}{10} = \frac{15}{40} = \frac{3}{8}$$

Step 1: Multiply numerators 3 x 5 = 15

Step 2: Multiply denominators 4 x 10 = 40

Step 3: Reduce fraction $\frac{15}{40} \div \frac{5}{5} = \frac{3}{8}$

$$\frac{2}{7} \times \frac{4}{12} = \frac{8}{84} = \frac{2}{21}$$

Practice 2.5

1) $\frac{6}{7}$ x $\frac{5}{13}$

2) $\frac{2}{3}$ x $\frac{7}{12}$

5) $\frac{4}{11}$ x $\frac{5}{8}$

6) $\frac{7}{10}$ x $\frac{1}{2}$

7) $\frac{6}{13}$ x $\frac{3}{5}$

3) $\frac{1}{4}$ x $\frac{1}{2}$

4) $\frac{3}{8}$ x $\frac{2}{3}$

8) $\frac{4}{9}$ x $\frac{1}{3}$

9) $\frac{7}{8}$ x $\frac{3}{4}$

10) $\frac{6}{11}$ x $\frac{4}{8}$

2.6 Dividing Fractions

To divide fractions you rewrite the first fraction (keep), change your operation sign from division to multiplication (Change), and take the reciprocal or flip your second fraction (Change), and finally reduce. This method is known to some as the Keep-Change-Change method.

Example

$$\frac{3}{4} \times \frac{5}{10} =$$

$$\frac{3}{4} \div \frac{10}{5} = \frac{30}{20} = 1\frac{10}{20} = 1\frac{1}{2}$$

Step 1: Keep – Rewrite the first fraction

Step 2: Change – Change the sign to division

Step 3: Change - Take the reciprocal of or flip the second fraction

Step 4: Reduce to lowest terms

Practice 2.6

1) $\frac{6}{7} \div \frac{5}{13}$

2) $\frac{5}{3} \div \frac{7}{15}$

5) $\frac{6}{11} \div \frac{5}{8}$

3) $\frac{1}{6} \div \frac{1}{5}$

4) $\frac{3}{8} \div \frac{5}{3}$

6) $\frac{7}{10} \div \frac{1}{5}$

7) $\frac{6}{13} \div \frac{3}{5}$

9) $\frac{7}{8} \div \frac{3}{6}$

8) $\frac{6}{9} \div \frac{1}{3}$

10) $\frac{6}{11} \div \frac{6}{8}$

2.7 Adding Fractions

In order to add fractions you must have the same denominators so we will start out with two fractions with the same denominator.

$$\frac{5}{9} + \frac{1}{9} = \frac{6}{9} = \frac{2}{3}$$

Step 1: Add your numerators 5 + 1 = 6

Step 2: The denominator stays the same

Step 3: Reduce if possible

If the denominators are different we must first find the least common denominator, find an equivalent fraction, add our fractions and then reduce.

$$\frac{5}{9} + \frac{2}{4} = \frac{20}{36} + \frac{18}{36} = \frac{38}{36} = 1\frac{2}{36} = 1\frac{1}{18}$$

Step 1: Find the least common denominator for 4 and 9 which is 36

Step 2: Find equivalent fractions using the new denominator of 36.

Step 3: Add the new equivalent fractions

Step 4: Reduce if possible

Practice 2.7

1) $\frac{2}{5} + \frac{1}{2}$

3) $\frac{6}{10} + \frac{2}{3}$

2) $\frac{2}{5} + \frac{3}{4}$

4) $\frac{3}{5} + \frac{1}{2}$

5) $\frac{4}{10} + \frac{1}{4}$

6) $\frac{2}{3} + \frac{3}{4}$

7) $\frac{1}{2} + \frac{2}{4}$

9) $\frac{1}{5} + \frac{1}{4}$

8) $\frac{1}{5} + \frac{2}{3}$

10) $\frac{8}{10} + \frac{3}{4}$

2.8 Subtracting Fractions

In order to subtract fractions you must have the same denominators so we will start out with two fractions with the same denominator.

$$\frac{5}{9} - \frac{1}{9} = \frac{4}{9}$$

Step 1: Subtract your numerators 5 - 1 = 6
Step 2: The denominator stays the same
Step 3: Reduce if possible

If the denominators are different we must first find the least common denominator, find an equivalent fraction, subtract our fractions and then reduce.

$$\frac{5}{9} - \frac{2}{4} = \frac{20}{36} - \frac{18}{36} = \frac{2}{36} = \frac{1}{18}$$

Step 1: Find the least common denominator for 4 and 9 which is 36
Step 2: Find equivalent fractions using the new denominator of 36.
Step 3: Subtract the new equivalent fractions
Step 4: Reduce if possible

Practice 2.8

1) $\frac{2}{5} - \frac{1}{2}$

3) $\frac{6}{10} - \frac{2}{3}$

2) $\frac{2}{5} - \frac{3}{4}$

4) $\frac{3}{5} - \frac{1}{2}$

6) $\frac{2}{3} - \frac{3}{4}$

5) $\frac{4}{10} - \frac{1}{4}$

7) $\frac{1}{2} - \frac{2}{4}$

9) $\dfrac{1}{5} - \dfrac{1}{4}$

8) $\dfrac{1}{5} - \dfrac{2}{3}$

10) $\dfrac{8}{10} - \dfrac{3}{4}$

2.9: Adding, Subtracting, Multiplying, Dividing Improper and Mixed Numbers

When adding and subtracting improper fractions we must go through the same steps as with proper fractions which is find an LCD, equivalent fractions, and then add or subtract. So let's look at a few examples:

$$\frac{12}{5} - \frac{4}{6} = \frac{72}{30} - \frac{20}{30} = \frac{52}{30} = 1\frac{22}{30} = 1\frac{11}{15}$$

$$\frac{10}{7} + \frac{1}{2} = \frac{20}{14} + \frac{7}{14} = \frac{27}{30} = \frac{9}{10}$$

When multiplying and dividing improper fractions you follow the same steps that you follow with common fractions so let's do a few of those:

$$\frac{6}{2} \times \frac{5}{9} = \frac{30}{18} = 1\frac{12}{18} = 1\frac{2}{3}$$

$$\frac{7}{3} \div \frac{14}{5} = \frac{7}{3} \times \frac{5}{14} = \frac{35}{42} = \frac{5}{6}$$

When adding and subtracting mixed numbers the first step is to change the mixed number into an improper fraction, find the LCD, equivalent fraction and add or subtract. So let's look at a few examples:

$$1\frac{2}{5} - 1\frac{1}{9} = \frac{7}{5} - \frac{10}{9} = \frac{63}{45} - \frac{50}{45} = \frac{13}{5} = 2\frac{3}{5}$$

$$2\frac{12}{15} + \frac{4}{6} = \frac{42}{15} - \frac{4}{6} = \frac{84}{30} - \frac{20}{30} = \frac{64}{30} = 2\frac{4}{30} = 2\frac{2}{15}$$

When multiplying and dividing mixed numbers change the mixed number to an improper fraction and then you follow the same steps that you follow with common fractions so let's do a few of those:

$$3\frac{1}{7} \times \frac{4}{6} = \frac{22}{7} \times \frac{4}{6} = \frac{88}{42} = 2\frac{4}{42} = 2\frac{2}{21}$$

$$2\frac{2}{8} \div \frac{3}{4} = \frac{18}{8} \div \frac{3}{4} = \frac{18}{8} \times \frac{4}{3} = \frac{72}{24} = 3$$

Chapter 2 Review:

Find the LCM and GCF of each set of numbers:

	LCM	GCF
1) 21, 30	_____	_____
2) 16, 4	_____	_____
3) 5, 4	_____	_____
4) 4, 34	_____	_____
5) 30, 4	_____	_____

Find the LCD for each number below:

6) $\frac{11}{13}$ and $\frac{6}{7}$

7) $\frac{2}{3}$ and $\frac{3}{10}$

8) $\frac{8}{15}$ and $\frac{3}{7}$

9) $\frac{8}{10}$ and $\frac{3}{5}$

10) $\frac{1}{9}$ and 8

Change the mixed fractions to improper and improper to mixed

11) $\frac{9}{4}$

12) $\frac{30}{7}$

13) $\frac{16}{5}$

14) $\frac{52}{13}$

15) $\frac{19}{12}$

16) $9\frac{1}{5}$

17) $6\frac{1}{2}$

18) $2\frac{7}{9}$

19) $7\frac{5}{8}$

20) $9\frac{11}{12}$

Solve each problem according to the operation listed:

21) $\frac{1}{5} \div \frac{5}{16}$

22) $\frac{1}{6} \times \frac{7}{12}$

23) $2\frac{8}{10} - 1\frac{1}{3}$

24) $1\frac{2}{9} \times \frac{4}{7}$

25) $\frac{12}{5} - \frac{1}{9}$

26) $\frac{17}{8} \times \frac{5}{3}$

27) $\frac{3}{4} - \frac{3}{16}$

28) $\frac{5}{16} + \frac{8}{14}$

29) $3\frac{2}{5} + \frac{1}{2}$

30) $2\frac{2}{5} \div \frac{1}{6}$

31) $\frac{15}{7} + \frac{1}{3}$

32. $\frac{21}{13} \div \frac{11}{6}$

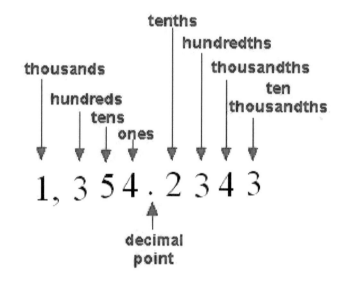

Chapter 3: Decimals

3.1 Introduction to Decimals

A decimal number is a number that contains a decimal point that goes between the tenths place and the unit or ones place. In order to understand decimals we must start with place value.

Place value is the position in which a number lies. Let's look at the place value chart below:

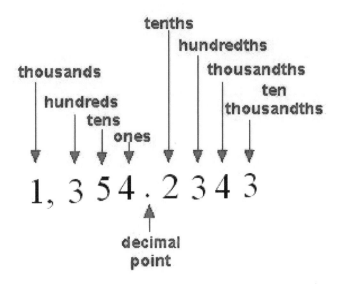

As you can see each number (digit) has a specific place value. The numbers that fall to the left of the decimal point are whole numbers and those to the right are decimal numbers (fractions). As we move to the left each number is ten times bigger and to the right ten times smaller.

Practice 3.1

What digit is listed in the positions below?

4,798,354,263.412569

1) Hundredths place   _____

2) Thousands place   _____

3) Ten-thousandths place   _____

4) Millions place   _____

5) Billions place   _____

6) Ones place   _____

7) Tenths place   _____

8) Thousandths place   _____

3.2 Adding Decimals

In order to add decimals you must line your numbers up according to the decimal point and then add. You can put zeroes in place of your blank spaces but it is not necessary. We will solve a problem with and without zeroes:

Add 1.632 to 1.5

$$
\begin{array}{r}
1.632 \\
+\ 1.500 \\
\hline
3.132
\end{array}
$$

$$
\begin{array}{r}
1.632 \\
+\ 1.5 \\
\hline
3.132
\end{array}
$$

Practice 3.2: Add

1) 6.3 + 2.2

2) 1.7 + 4.3

3) 9.1 + 5.4

4) 0.6 + 2.7

5) 7.6 + 1.3

6) 6.7 + 9.6

7) 3.7 + 8.6

8) 5.3 + 2.6

9) 91.1 + 4.6

10) 54.6 + 6.6

3.3 Subtracting Decimals

In order to subtract decimals you must line your numbers up according to the decimal point and then subtract. You can put zeroes in place of your blank spaces but it is not necessary. We will solve a problem with and without zeroes:

$$
\begin{array}{r}
2.894 \\
+\ 1.720 \\
\hline
1.174
\end{array}
$$

$$
\begin{array}{r}
2.894 \\
+\ 1.72 \\
\hline
1.174
\end{array}
$$

Practice 3.3: Subtract

1) 63.547 − 12.004   6) 62.547 − 54.763

2) 21.084 − 1.078   7) 35.453 − 13.4

3) 74.547 − 5.894   8) 63.882 − 6.487

4) 83.035 − 46.089   9) 89.547 − 16.59

5) 99.784 − 80.499   10) 117.536 − 11.234

3.4 Multiplication of Decimals

In order to multiply decimals you must place your longest number on the top and multiply as if the decimal point(s) are not there. Once you have completed the multiplication process go back to your problem and count the number of digits behind the decimal point(s) and move your decimal point the same number of times in your product (answer).

For example:

75.853 × 234.12

```
      75.853
×     234.12
     151706
     758530
   30341200
  227559000
 1517060000
17758.70436
```

Both of our numbers are the same number of digits so we can place either one on top.

We multiply as if the decimal points are not there and we get a product of 1775870436.

We count the number of places behind the decimal point in our original problem which is five.

Starting from the end of our product (the right side) we move our decimal point over five places and get a final answer of 17758.70436

Practice 3.4

1) 74.38 × 4.8   4) 86.3 × 9.0 0   7) 96.68 × 2.1

2) 18.81 × 2.41   5) 9.98 × 2.8   8) 96.2 × 5.5

3) 17.7 × 6.38   6) 11.19 × 8.8   9) 46.8 × 7.86

3.5 Division of Decimals

Before we discuss dividing decimals let's go back and talk about the parts of a division problem.

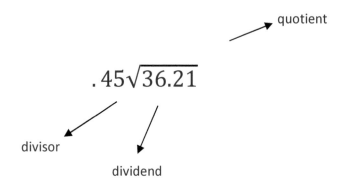

quotient

$$.45\overline{)36.21}$$

divisor

dividend

In order to divide decimals you must make your divisor a whole number by moving the decimal point to the right the number of spaces necessary to make it a whole number so in this problem we must move over one space. What you do to the divisor you must also do to the dividend. So the decimal in the dividend must be moved over one space as well. That leaves us with the following:

$$45.\overline{)3621.}$$

Bring the decimal point in your dividend up to your quotient and solve. Since our decimal point is at the end there is no need to move it up to the quotient however in normal circumstances if it is not at the end we will move it up into the quotient.

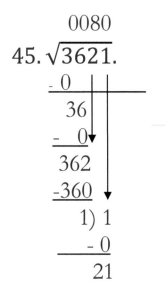

Practice 3.5 (Round to the nearest hundredth)

1) $.23\overline{)9581}$

2) $.82\overline{)1883}$

3) $4.6\sqrt{266.3}$

4) $1.4\sqrt{956.8}$

5) $5.0\sqrt{6624}$

6) $6.0\sqrt{83.65}$

7) $.19\sqrt{3.562}$

8) $1.5\sqrt{9.970}$

9) $1.6\sqrt{61.48}$

10) $3.5\sqrt{70.00}$

Chapter 3 Review

What digit is listed in the positions below?

$$523{,}475{,}896{,}145.236598740$$

1) Hundredths place   _____

2) Thousands place   _____

3) Ten-thousandths place   _____

4) Millions place   _____

5) Billions place   _____

6) Ones place   _____

7) Tenths place   _____

8) Thousandths place   _____

Solve and round to the nearest hundredth.

9) $7.5 + 33.92$

10) $2.89 + 56.4$

11) $75.8 + 61.74$

12) $0.8 + 2.337$

13) $96.84 + 17.614$

14) $78.526 - 17.8543$

15) $22.9856 - 3$

16) $16.713 - 7.91$

17) $856.29 - 367.41$

18) $97.832 - 71.85$

19) 824.87×3.9

20) 172.42×5.67

21) $70.8 \times 7.2 \ 1416$

22) $9.2 \times 6.6 \ 552$

23) 66.91×7.33

24) $.762\sqrt{37.34}$

27) $.1452\sqrt{28.6324}$

25) $5.96\sqrt{127.841}$

28) $9.7\sqrt{5.524}$

26) $3.72\sqrt{7245}$

Proportion Example

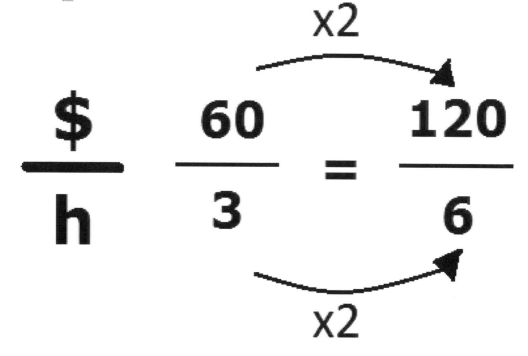

$$\frac{\$}{h} \qquad \frac{60}{3} = \frac{120}{6}$$

x2 (top)
x2 (bottom)

Chapter 4: Ratios & Proportions

4.1 Ratios

A ratio is a comparison between two different things. For example, you might want to find out what's the ratio of girls to boys in a classroom. Let's say that there are 15 girls and 20 boys in Mrs. Smith's classroom. We signify that this is a ratio by placing a colon between the two numbers.

15:20

When working with ratios you want to make sure you write the numbers in the order that they are given otherwise you will have an improper ratio. Ratios can also be written as fractions:

$$\frac{15}{20}$$

Or you can use the word "to" in between the numbers instead of using a color:

15 to 20

A ratio is not something that you can solve, at most you can reduce a ratio but there is no "solving" involved.

So the above ratio 15 to 20 can be reduced to 3 to 4 (3:4) because both numbers can be divided equally by 5.

Practice 4.1: Reduce each ratio if possible

   1) 15:90

   2) 13:52

   3) 16:64

   4) 1:3

   5) 4:9

   6) 6:36

   7) 14:42

   8) 50:250

   9) 15:225

   10) 17:34

4.2 Rates

A **rate** is a comparison of two quantities that have different units. Rates are always written as fractions.

Example: A distance runner ran 28 miles in 4 hours. The distance to time rate is:

28 miles / 4 hours = 14 miles / 2 hours = 7miles / 1 hour

A **unit rate**, is a rate that has a 1 in the denominator

Example: $8.75 / 1 pound or $8.75/pound is read as "$8.75 per pound"

On a trip, I traveled 344 miles before my car ran out of gas. My tank holds 18 gallons of gas. What is the unit rate that I traveled?

276 miles/18 gallons = 15.33 miles/gallon

Divide the number in the numerator by the number in the denominator of the rate.

Practice 4.2: Write as a rate and unit rate, round your answer to the nearest hundredth

	Rate	Unit Rate
1) 8 pencils for 13 dollars	_____	_____
2) 145 miles on 5 gallons of gas	_____	_____
3) 8 inches of snow in 7 hours	_____	_____
4) Mowed 3 yards for $35.00	_____	_____
5) 6 calculators cost $200.00	_____	_____
6) 8 chocolate bars cost 21 dollars	_____	_____
7) 9 dollars for 4 cans of tuna	_____	_____
8) 20 dollars for 4 books	_____	_____
9) 7 movie tickets cost $30.00	_____	_____
10) 8 batteries cost 28 dollars	_____	_____

4.3 Proportions

A **proportion** is an expression of the equality of two ratios or rates. In order to solve a proportion you cross multiply.

$$\frac{50}{6} \bowtie \frac{x}{3}$$

Multiply diagonally (cross multiply) where $50 \times 3 = 6 \times x$

Divide both sides by 6 to solve for x.

Plug x into original problem.

$$150 = 6x$$

$$\frac{150}{6} = \frac{6x}{6}$$

$$25 = x$$

$$\frac{50}{6} = \frac{25}{3}$$

Practice 4.3

1) $\frac{d}{63} = \frac{75}{98}$

2) $\frac{x}{30} = \frac{80}{50}$

3) $\frac{d}{53} = \frac{52}{15}$

4) $\frac{c}{87} = \frac{78}{13}$

5) $\frac{49}{62} = \frac{25}{c}$

6) $\frac{49}{27} = \frac{19}{d}$

7) A boat travels 42 miles in 3 hours (with a constant speed). How much time will it take traveling 273 miles?

8) 9 kg of apples cost $36. How many kilograms of apples can you get with $200?

9) A car travels 168 miles in 5 hours (with a constant speed). How much time will it take traveling 336 miles?

10) A boat can travel 273 miles on 39 gallons of gasoline. How much gasoline will it need to go 140 miles?

Chapter 4 Review

Solve each proportion and round answers to the nearest hundredth:

1) $\frac{65}{82} = \frac{x}{67}$

2) $\frac{71}{83} = \frac{12}{a}$

3) $\frac{94}{92} = \frac{x}{20}$

4) $\frac{8}{d} = \frac{11}{33}$

5) $\frac{57}{13} = \frac{34}{c}$

6) $\frac{7}{a} = \frac{47}{99}$

7) $2:4 =$ _____

8) $6:9 =$ _____

9) $5:15 =$ _____

10) $12:20 =$ _____

11) $18:45 =$ _____

12) 21:49 = _____

13) 33:121 = _____

14) 30:72 = _____

15) 39:52 = _____

Solve for x

16) x:3 = 2:6

17) 4:2 = x:1

18) 3:x = 2:4

19) 9:3 = 3:x

20) 7:2 = x:4

21) Mike bought 5 oranges for $1.30. What is the unit rate?

22) Amanda spent 80 minutes typing a 4,000 word essay. What's her speed in words per minute?

23) The copier company supplies 18 reams of paper for $55.84. What is the unit price per ream?

100%

$$\frac{1}{2} = 2\sqrt{\overset{\displaystyle 0.50}{1.00}}$$

Chapter 5: Percent

5.1 Percent, Fractions, Decimals

Percent means so many per 100 and is denoted by a percent sign %. So 50% means 50 per 100 or 50 out of 100 which is also the same as $\frac{50}{100} = \frac{1}{2}$.

To convert a decimal to a percent you simply move your decimal point over two places to the right and add a percent sign.

$$\text{Ex. } 0.3965 = 39.65\%$$

Practice

1) Express 0.0897 as a percent.

2) Express 0.625 as a percent.

3) Express 2.995 as a percent.

4) Express 32.8695 as a percent.

5) Express 0.014562 as a percent.

6) Express 161.5698 as a percent.

7) Express .0895 as a percent.

8) Express .8967 as a percent.

9) Express .245 as a percent.

10) Express 16.968 as a percent.

To convert a percent to a decimal move your decimal point two places to the left and drop the percent sign.

$$\text{Ex. } 29.8\% = .298$$

Practice

1) Express 89.6% as a decimal.

2) Express 13.268% as a decimal.

3) Express 109.5% as a decimal.

4) Express .95475% as a decimal.

5) Express 0.8965% as a decimal.

6) Express 741.24% as a decimal.

7) Express 18.5% as a decimal.

8) Express 63.568% as a decimal.

9) Express .06352% as a decimal.

10) Express 1.52% as a decimal.

To convert a fraction to decimal you divide the numerator by the denominator. $\frac{numerator}{denominator} = \frac{18}{24}$.

$$\text{Ex. } \frac{18}{24} = 24\overline{)18.00} \begin{array}{r} .75 \\ \hline -168 \\ \hline 120 \\ -120 \\ \hline 0 \end{array}$$

$$so \ \frac{18}{24} = .75$$

Practice – Round to the nearest hundredth.

1) Convert $\frac{2}{9}$ to a decimal.

2) Convert $\frac{5}{4}$ to a decimal.

3) Convert $\frac{3}{21}$ to a decimal.

4) Convert $\frac{12}{17}$ to a decimal.

5) Convert $\frac{10}{100}$ to a decimal.

6) Convert $\frac{13}{52}$ to a decimal.

7) Convert $\frac{11}{6}$ to a decimal.

8) Convert $\frac{65}{98}$ to a decimal.

9) Convert $\frac{23}{63}$ to a decimal.

10) Convert $\frac{19}{38}$ to a decimal.

To convert a decimal to a fraction take the digits behind the decimal point and make that the numerator. Your denominator is determined by the position in which your decimal ends for example if it stops in the hundredths place your denominator will be 100. Once you have the numerator and denominator reduce the fraction if possible.

<u>Let's look at an example:</u>

$$.85 = \frac{85}{100}$$

This fraction can be reduced by dividing both the numerator and denominator by 5

$$\frac{85}{100} = \frac{17}{20}$$

Practice

1) Convert .75 to a fraction.

2) Convert 25.6 to a fraction.

3) Convert 894.5 to a fraction.

4) Convert 13.7 to a fraction.

5) Convert 96.81 to a fraction.

6) Convert .745 to a fraction.

7) Convert .6352 to a fraction.

8) Convert .45 to a fraction.

9) Convert 1.42 to a fraction.

10) Convert 7.653 to a fraction.

To convert a fraction to a percent change the fraction to a decimal and change the decimal to percent by moving the decimal point two places to the right and adding a percent sign.

Example:
$$\frac{25}{40} = 40\overline{\smash{\big)}25.000}$$

$$
\begin{array}{r}
.625 \\
40\overline{)25.000} \\
\underline{-240} \\
100 \\
\underline{-80} \\
200 \\
\underline{-200} \\
00
\end{array}
$$

$.625 = 62.5\%$

Practice

1) Covert $\frac{2}{8}$ to a percent.

2) Covert $\frac{3}{7}$ to a percent.

3) Covert $\frac{13}{25}$ to a percent.

4) Covert $\frac{14}{57}$ to a percent.

5) Covert $\frac{45}{180}$ to a percent.

6) Covert $\frac{96}{108}$ to a percent.

7) Covert $\frac{17}{34}$ to a percent.

8) Covert $\frac{49}{63}$ to a percent.

9) Covert $\frac{21}{28}$ to a percent.

10) Covert $\frac{15}{45}$ to a percent.

To convert a percent to a fraction convert the percent to a decimal first and then convert the decimal to a fraction reduced to lowest terms.

$$37.8\% = .378 = \frac{378}{1000} = \frac{189}{500}$$

Practice

1) Convert 47% to a fraction.

2) Convert 1.63% to a fraction.

3) Convert 12.74% to a fraction.

4) Convert 93.8% to a fraction.

5) Convert 89% to a fraction.

6) Convert 107% to a fraction.

7) Convert 16% to a fraction.

8) Convert 52% to a fraction.

9) Convert 45.8% to a fraction.

10) Convert 72% to a fraction.

5.2 Rate, Base, Part

In percent problems there are three key pieces of information that must be identified in order to solve the problem. You must be able to identify the Rate, Base, and Part (Portion) in order to solve a percent problem.

Rate (R) = the percentage (this is the number with the percent sign).

Base (B) = the whole amount and typically follows the word "of" in the problem.

The Part or Portion (P) = part of the whole (the "is").

The formulas used to solve percentage problems are:

$$R = \frac{P}{B} \qquad\qquad P = BR \qquad\qquad B = \frac{P}{R}$$

The diagram below will help you remember the formulas:

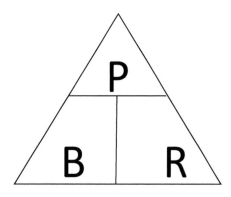

Example 1: Identify the Rate, Base, and Part and solve.

What is 30% of 24?

Rate = 30%

Part = unknown

Base = 24

So we would use P=BR to solve; P= (24)(30%) = 7.2

Example 2: Identify the Rate, Base, and Part and solve.

12 is 36% of what number?

Part = 12

Rate = 36%

Base = unknown

So we would use $B = \dfrac{P}{R} = \dfrac{12}{36\%} = 33.\overline{3}$

Example 3: Identify the Rate, Base, and Part and solve.

5 is what percent of 19?

Part = 5

Rate = unknown

Base = 19

So we would use $R = \dfrac{P}{B} = \dfrac{5}{19} = .2632 = 26.32\%$

Practice 5.2 (Round to the nearest hundredth):

1) What is 25% of 350?

2) 15% of what number is 95?

3) 30 is what percent of 90?

4) 75 is what percent of 850?

5) 10% of what number is 245?

6) 16% what number is 85?

7) What percent of 120 is 30?

8) Jan has a 15% commission rate on all her sales. If she sells$ 60,000 worth of merchandise in 1 month, what commission will she earn?

9) 29% of Kanika's monthly salary is deducted for withholding. If those deductions total $408, what is her salary?

10) In a chemistry class of 28 students, 7 received a grade of A. What percent of the students received A's?

5.3 Percent Problems: Proportion Method

When using the proportion method you use the same formula however you will plug in the numbers that are given to you in the problem so your unknown will change for each type of problem.

$$\frac{\text{part}}{\text{whole}} = \frac{\text{percent}}{100}$$

Example 1: What is 30% of 24?

In this problem the part is unknown so we label it as x.

$$\frac{x}{24} \diagup\!\!\!\!\diagdown \frac{30}{100}$$

Cross multiply and solve for your unknown (x).

$$100x = (30)(24)$$

$$100x = 720$$

$$\frac{100x}{100} = \frac{720}{100}$$

$$x = 7.2$$

Example 2:

12 is 36% of what number?

$$\frac{12}{x} \diagup\!\!\!\!\diagdown \frac{36}{100}$$

$$1200 = 36x$$

$$\frac{36x}{36} = \frac{1200}{36}$$

$$x = 33.\overline{3}$$

Example 3:

5 is what percent of 19?

$$\frac{5}{19} \diagdown \frac{x}{100}$$

$$500 = 19x$$

$$\frac{500}{19} = \frac{19x}{19}$$

$$x = 26.32\%$$

Practice 5.3

1) What is 20% of 60?

2) 12 is 75% of what number?

3) 6 is what percent of 8?

4) 8 is 40% of what number?

5) 33.3% of what nmber is 24?

6) Find 12.5% of 400.

7) What is $\frac{5}{8}$ of 32?

8) 12 is $\frac{2}{5}$ of what number?

9) What is 87.5% of 150?

10) 25 is what percent of 30?

Chapter 5 Review

1) Express 12.365 as a percent.

2) Express 0.9856 as a percent.

3) Express 7.8596 as a percent.

4) Express 21.45 as a percent.

5) Express 0.04859 as a percent.

6) Express 13.8% as a decimal.

7) Express 61.87% as a decimal.

8) Express 125.75% as a decimal.

9) Express 0.87512% as a decimal.

10) Express 97% as a decimal

11) Convert $\frac{3}{9}$ to a decimal.

12) Convert $\frac{5}{15}$ to a decimal.

13) Convert $\frac{3}{45}$ to a decimal.

14) Convert $\frac{19}{17}$ to a decimal.

15) Convert $\frac{10}{36}$ to a decimal.

16) Convert 0.41 to a fraction.

17) Convert 18.9 to a fraction.

18) Convert 106.5 to a fraction.

19) Convert 24.6 to a fraction.

20) Convert 97.65 to a fraction.

21) Covert $\frac{3}{8}$ to a percent.

22) Covert $\frac{8}{7}$ to a percent.

23) Covert $\frac{13}{26}$ to a percent.

24) Covert $\frac{41}{57}$ to a percent.

25) Covert $\frac{45}{75}$ to a percent.

26) Convert 32% to a fraction.

27) Convert 12.45% to a fraction.

28) Convert 3.19% to a fraction.

29) Convert 87.62% to a fraction.

30) Convert 57% to a fraction.

Solve using the Rate, Part, Base Formulas

31) What is 13% of 69?

32) 15 is 65% of what number?

33) 9 is what percent of 81?

34) 7 is 32% of what number?

35) 16.7% of what nmber is 36?

Solve using the Proportion Method

36) Find 17.5% of 550.

37) What is $\frac{3}{8}$ of 29?

38) 10 is $\frac{4}{5}$ of what number?

39) What is 36.5% of 75?

40) 49 is what percent of 85?

Cumulative Review: Chapters 1-5

Solve each problem using the specified operation:

1) 93 + 128

2) 65 + 39

3) 1938 + 457

4) 394 - 198

5) 182 – 57

6) 275 – 194

7) 72 x 81

8) 194 x 26

9) 37 x 18

10) $15\sqrt{145}$

11) $17\sqrt{134}$

12) $81\sqrt{386}$

Find the Prime Factorization

13) 192

14) 235

15) 769

Find the LCM and GCF of each set of numbers:

16) 36, 48

17) 17, 51

18) 3, 123

Find the LCD for each set of numbers below

19) $\frac{9}{17}$ and $\frac{3}{5}$

20) $\frac{5}{8}$ and $\frac{2}{3}$

21) $\frac{1}{15}$ and $\frac{4}{7}$

Change the mixed fractions to improper and the improper to mixed

22) $\frac{19}{4}$

23) $3\frac{7}{13}$

24) $\frac{57}{8}$

25) $11\frac{2}{5}$

Solve each problem according to the operation listed and reduce to lowest terms:

26) $\frac{3}{7} + \frac{5}{9}$

27) $1\frac{2}{8} \ x \ \frac{1}{4}$

28) $\frac{7}{17} \div \frac{4}{5}$

29) $8\frac{1}{8} - 3\frac{5}{6}$

30) $56.523 - 1.745$

31) $19.4 + 3.89$

32) $4.17 \ x \ 2.6$

33) $63.745 \div .352$

Solve each proportion and round answers to the nearest hundredth

34) $9:27$

35) $13:39$

36) $\frac{5}{7} = \frac{10}{x}$

37) $\frac{6}{8} = \frac{x}{9}$

Express as a fraction and decimal

38) 45%

39) 9.75%

40) 367.2%

Express as a percent and a decimal

41) $\frac{5}{9}$

42) $\frac{7}{17}$

Solve

43) What is 24% of 32?

44) 17 is what percent of 138?

45) 3% of 87 is what number?

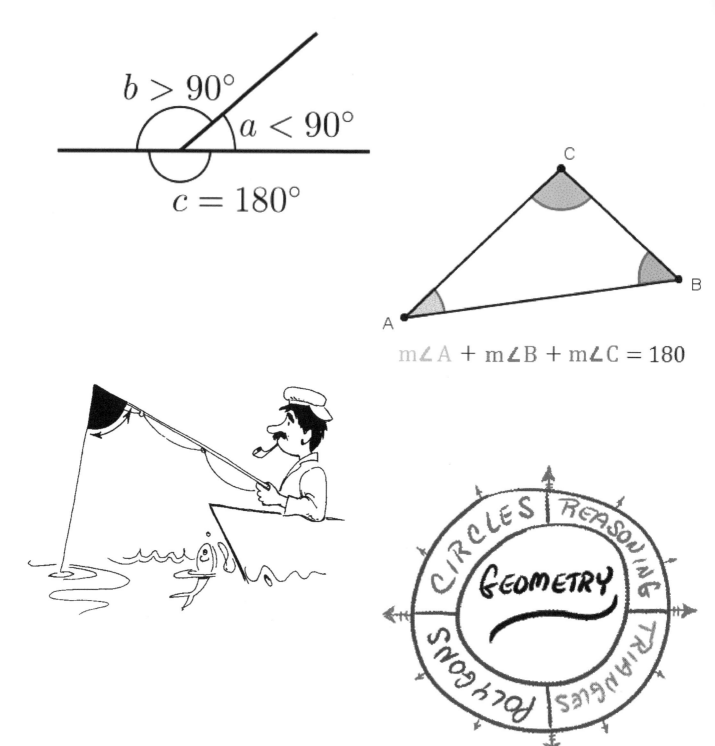

$b > 90°$

$a < 90°$

$c = 180°$

$m\angle A + m\angle B + m\angle C = 180$

Chapter 6: Geometry

Chapter 6: Geometry

6.1 Types of Angles

Straight Angles

A straight angle is exactly what it states. It is an angle that is a straight line that measures exactly 180°.

Right Angles

Right angles are angles that measure exactly 90°.

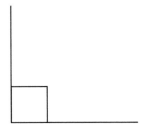

Obtuse Angles

Obtuse angles are those which are greater than 90° but less than 180°, that is, $90° < \Theta < 180°$. Any angle between the dotted lines is an obtuse angle.

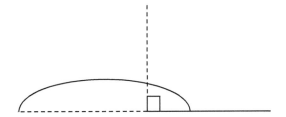

Acute Angles

An acute angle is an angle that measures between 0° and 90

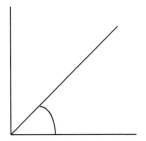

Adjacent Angles

Two angles are adjacent when they have a common side and a common vertex but do not overlap.

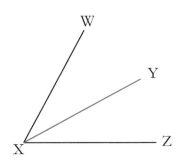

<WXY is adjacent to <YXZ because the share a common side XY and a common vertex X

Complementary Angles

Angles are complimentary when they add up to 90°. (Complimentary angles DO NOT have to be together!)

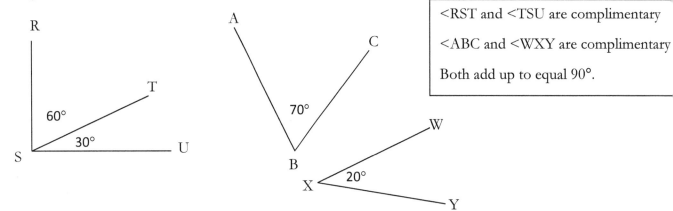

<RST and <TSU are complimentary

<ABC and <WXY are complimentary

Both add up to equal 90°.

Supplementary Angles

Angles are supplementary when they add up to 180°. (Supplementary angles DO NOT have to be together!)

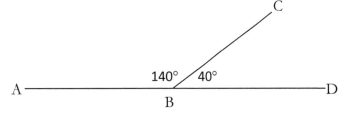

<ABC and <CBD are supplementary because they add up to 180°

<EFG and <HIJ are supplementary because they also add up to equal 180°

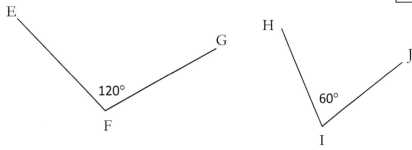

Vertical Opposite Angles

Vertical opposite angles are angles that are opposite to each other when two straight lines intersect.

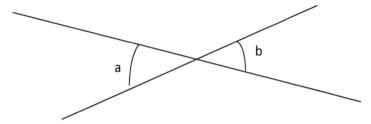

So <a and <b are vertical opposite angles and they are equal to each other.

Corresponding Angles

When two parallel lines are crossed by a line called the transversal, the angles formed which are in corresponding positions or the same positions, are called corresponding angles. Corresponding angles are equal to one other.

<a = <e

<b = <f

<c = <g

<d = <h

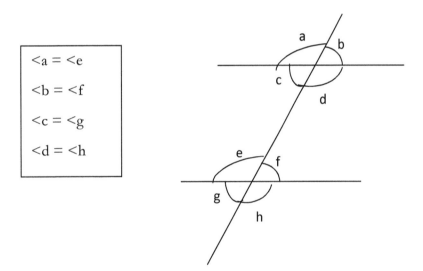

Practice 6.1

Classify each angle as obtuse, right, or acute.

1)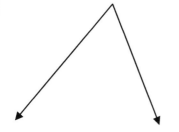

2)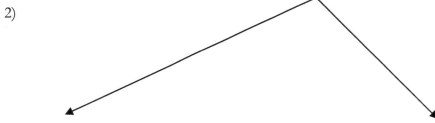

3)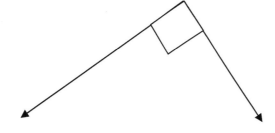

4)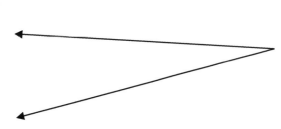

5) 76°

6) 139°

7) 174°

8) 96.8°

Find the compliment and supplement to each angle listed below:

9) 46°

10) 13.5°

11) 26.9°

12) 67°

6.2 Triangles

A triangle has three sides and three angles. The three angles should always add up to equal 180°

Equilateral Triangles

Triangles with all three sides equal in length and all three angles equal in measure, are called equilateral triangles. Since the angles in a triangle add up to 180° and the size of each angle is the same in an equilateral triangle, the angles are all 60°.

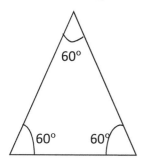

Isosceles Triangles

Isosceles triangles are triangles with two equal sides and two equal angles.

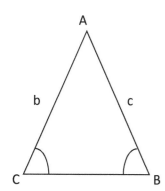

$$b = c$$
$$<B = <C$$

Scalene Triangles

A scalene triangle is a triangle that has no equal sides and no equal angles.

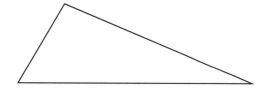

Right Triangles

Right triangles are triangles in which one of the angles is equal to 90°.

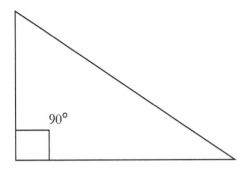

Obtuse Triangles

Obtuse triangles are triangles in which one of the angles is an obtuse angle so it is greater than 90° but less than 180°.

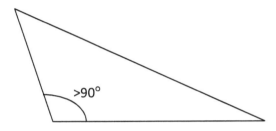

Acute Triangles

Acute triangles are triangles in which all the angles are acute or between 0° and 90°.

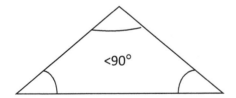

Practice 6.2

Identify each triangle and the number of angles that are the same within them.

1.

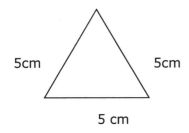

5cm   5cm

5 cm

2.

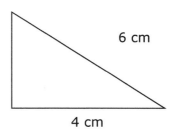

4 cm   6 cm

4 cm

3.

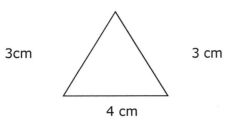

3cm   3 cm

4 cm

4.

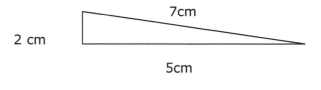

7cm

2 cm

5cm

Identify the missing angles and state whether the angles are acute or obtuse.

1.

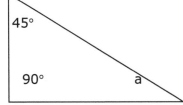

45°

90°   a

2.

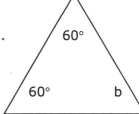

60°

60°   b

3.

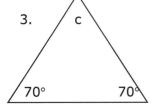

c

70°   70°

4.

d

90°   50°

6.3 Pythagorean Theorem

The Pythagorean Theorem is used to find the length of an unknown side in a right triangle. In order to solve using the Pythagorean Theorem you must have two of the sides so that you can find the third using the following formula.

$a^2 + b^2 = c^2$, where a, b, and c are the sides of the triangle.

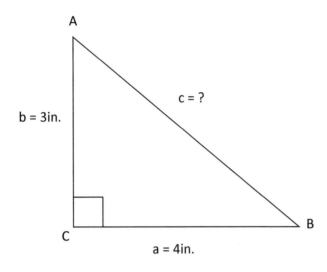

If a = 4in. and b = 3in. we can use the Pythagorean Theorem to find c.

$$a^2 + b^2 = c^2$$

$$4^2 + 3^2 = c^2$$

$$16 + 9 = c^2$$

$$25 = c^2$$

$$\sqrt{25} = c$$

$$5 = c$$

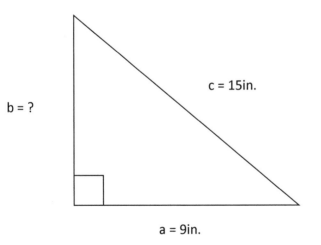

If a = 9in. and c = 15in. we can use the Pythagorean Theorem to find a.

$$a^2 + b^2 = c^2$$
$$\underline{-a^2 \qquad\qquad -a^2}$$
$$b^2 = c^2 - a^2$$

$$b^2 = 15^2 - 9^2$$

$$b^2 = 225 - 81$$

$$b^2 = 144$$

$$b = \sqrt{144}$$

$$b = 12$$

Practice 6.3

Solve using the Pythagorean Theorem:

1) a = 4, b = 3, c = ?
2) a = 6 ; b = ? ; c = 16;
3) a = 4 ; b = 2 ; c = ?
4) a = ? ; b = 3.5; c = 4.2

5)

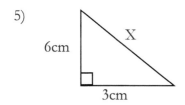

6)

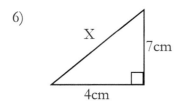

6.4 Quadrilaterals

Quadrilaterals are any four sided shapes.

Properties of Quadrilaterals:

- Four sides
- Four vertices
- Interior angles add up to $360°$

Types of Quadrilaterals

Rectangle

A rectangle is a four-sided figure that has four right angles and opposite sides are equal.

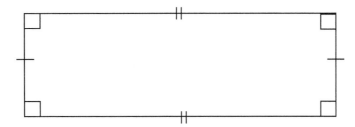

A rhombus is a four-sided figure in which all four sides are equal in length.
- Opposite sides are parallel *and* opposite angles are equal.
- A rhombus is also called a diamond

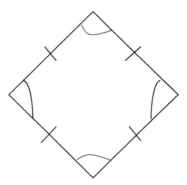

Square

A square is a four-sided figure with the following characteristics:

- All four sides are of equal length
- All four angles are $90°$

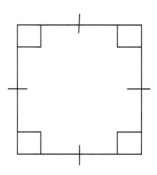

Parallelograms

A parallelogram has the following characteristics:

- Opposite sides are parallel and equal in length
- Opposite angles are equal

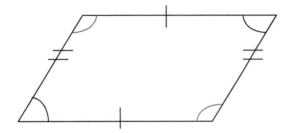

TIP: All rectangles, squares, and rhombuses are parallelograms!!!!

Chapter 6 Review

Finding the Long Side of a Right Angled Triangle ($S^2 + M^2 = L^2$) Problems 1-3
Round to the Nearest Hundredth

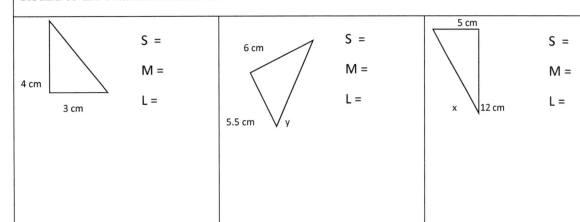

S =

M =

L =

S =

M =

L =

S =

M =

L =

Finding the Short Side of a Right Angled Triangle ($S^2 + M^2 = L^2$) Problems 4-6
Round to the Nearest Hundredth

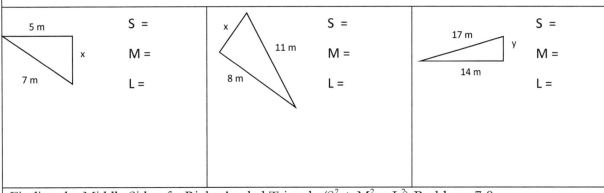

S =

M =

L =

S =

M =

L =

S =

M =

L =

Finding the Middle Side of a Right Angled Triangle ($S^2 + M^2 = L^2$) Problems 7-9

Round to the Nearest Hundredth

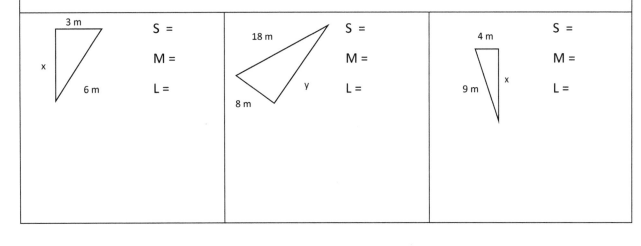

S =

M =

L =

S =

M =

L =

S =

M =

L =

10) Classify △*DEF* as equilateral, isosceles, or scalene.

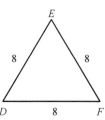

11) Name a right triangle.

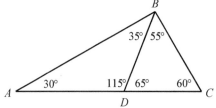

12) Name an obtuse triangle.

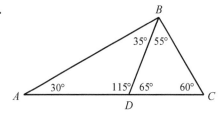

14) Name an acute triangle.

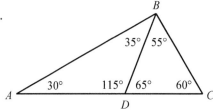

15) Name a right triangle.

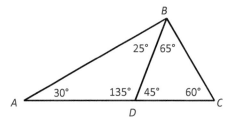

16) Draw and identify a triangle with angles measures of 60°, 60°, and 60°.

17) Draw and identify a triangle with angle measures of 45°, 45°, and 90°.

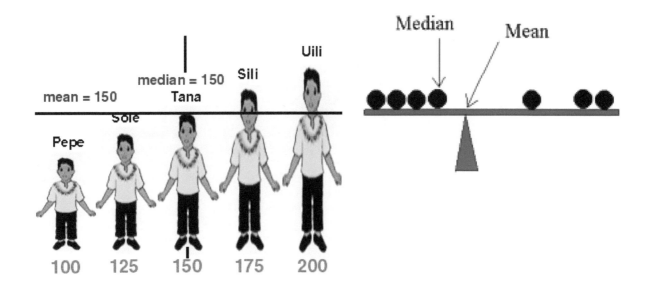

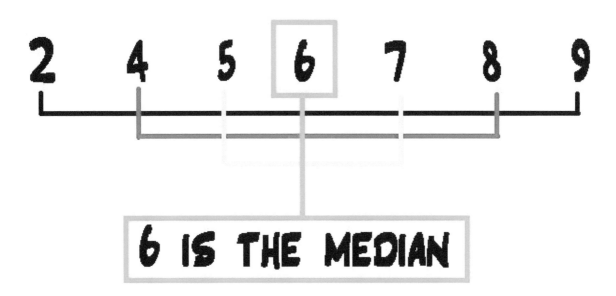

Chapter 7: Statistics

Chapter 7: Statistics

7.1 Mean

The mean of a set of numbers is the average of the numbers. In order to find the average take the total of the numbers and divide it by the amount of numbers that you are given.

$$\text{Mean} = \frac{the\ sum\ or\ total\ of\ the\ numbers}{the\ amount\ of\ numbers}$$

Example:

Find the mean of the following numbers:

7, 11, 12, 15, 25, 6, 8, 10, 24, 13

First find the total of the numbers

Divide by the amount of numbers given which is 10.

Mean = 13.1

7+ 11+ 12+ 15+ 25+ 6+ 8+ 10+ 24+ 13 = 131

$$\frac{131}{10} = 13.1$$

Practice 7.1

Find the mean of each of the following (Round to the nearest hundredth):

1) 17, 25, 14, 18, 9

3) 13, 18, 94, 56, 45, 32, 7, 9

2) 16, 1, 13, 3, 14, 75, 84

4) 3, 2, 5, 6, 4, 8, 7, 1, 19

5) 75, 80, 32, 90, 95, 81, 87, 77, 79, 63

6) 9, 15, 11, 12, 3, 5, 10, 20, 14, 6, 8, 8, 12, 12, 18, 15, 6, 9, 18, 11

7) 6, 11, 7

8) 3, 7, 5, 13, 20, 23, 39, 23, 40, 23, 14, 12, 56, 23, 29

9) 3, 7, 5, 13, 2

10) 8, 9, 13, 18

7.2 Median

The median of a set of numbers is the middle number when the numbers are placed in ascending order (order from least to greatest).
Example:
122, 130, 128, 123, 126, 124, 127, 125, 129

Placing in ascending order we get
122, 123, 124, 125, **126**, 127, 128, 129, 130
The middle number is 126 so 126 is the median

If there is an even amount of numbers, take the two middle numbers and find the average, this will be your median.

401, 406, 403, 405, 402, 404
Place in ascending order:
401, 402, 403, 404, 405, 406

There is an even set of numbers so there will be two middle numbers:
401, 402, **403, 404**, 405, 406
$403 + 404 = 807$
$$\frac{807}{2} = 403.5$$

So the median is **403.5**

Practice 7.2

Find the median of the numbers below:

1) 17, 25, 14, 18, 9

2) 16, 1, 13, 3, 14, 75, 84

3) 13, 18, 94, 56, 45, 32, 7, 9

4) 3, 2, 5, 6, 4, 8, 7, 1, 19

5) 75, 80, 32, 90, 95, 81, 87, 77, 79, 63

7.3 Mode

The mode is the most frequent number in a given set of numbers. So this is the number that is listed the most times.

Example:
Identify the mode:
20, 24, 27, 24, 26, 28, 27, 24

The number that appears most above is 24
20, **24**, 27, **24**, 26, 28, 27, **24**

TIP: There can be more than one Mode or there may not be one at all.

Practice 7.3

1) 4 , 3 , 3 , 7 , 3 , 9 , 3 , 8

2) 1 , 7 , 7 , 9 , 2 , 4 , 5 , 5

3) 2 , 5 , 8 , 8 , 7

4) 9 , 8 , 4 , 7 , 4 , 5 , 5 , 6

5) 2 , 3 , 8 , 3 , 6 , 2

7.4 Range

The range of a set of numbers is the difference between the largest and the smallest number.

Example:

Calculate the range of the following numbers:

204, 210, 215, 220, 225, 234, 238, 240

The range:

= the largest number – the smallest number

= 240 – 204

= 36

Practice 7.4

Find the Range

1) 3 , 3 , 5 , 3 , 6

2) 6 , 4 , 6 , 8 , 8 , 4 , 9 , 6 , 3

3) 9 , 4 , 4 , 8 , 2 , 4 , 4

4) 7 , 8 , 8 , 3 , 3 , 4 , 2

5) 2 , 14 , 5 , 9 , 9 , 9 , 8 , 2

6) 9 , 9 , 9 , 12 , 8 , 4 , 8

7) 8 , 6 , 6 , 3 , 3 , 19 , 3 , 2

8) 2 , 1 , 2 , 5 , 15 , 6 , 2 , 9

9) 2 , 9 , 5 , 18 , 2 , 2 , 2 , 2

10) 13 , 6 , 5 , 9 , 2

Chapter 7 Review

Find the mean, median, mode, and range for each of the problems below:

1) 18, 24, 17, 19, 24, 16, 22, 18

Mean_____
Median_____
Mode_____
Range _____

2) 75, 87, 49, 68, 75, 84, 98

Mean_____
Median_____
Mode_____
Range _____

3) 55, 47, 38, 66, 56, 64, 44, 63, 39

Mean_____
Median_____
Mode_____
Range _____

4) 25, 48, 25, 33, 57, 50

Mean_____
Median_____
Mode_____
Range _____

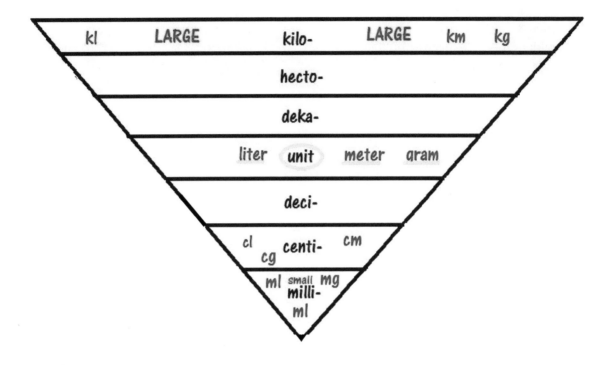

Chapter 8: Measurement

8.1 Converting Units

To convert units use the conversion chart below:

1 inch = 2.54 centimeters
1 meter = 3.2808399 feet
1 meter = 1.0936133 yard
1 mile = 1.609344 kilometer
1 kilogram = 2.20462262 pounds
1 acre = 43,560 sq feet
1 acre = 40.4685642 are
1 are = 100 square meters

1 hectare = 100 are = 2.47105381 acre
1 US fluid ounce = 29.5735296 mL
1 liter = 1.05668821 US quarts
1 US gallon = 3.78541178 liter
1 meter / second = 2.23693629 mph
1 meter / second = 3.2808399 feet / second
1 mph = 1.609344 kph

Temperature

$$°F = °C \times \frac{9}{5} + 32° F$$

$$°C = (°F - 32) \times \frac{5}{9}$$

Converting Units within SI system

Length
1cm = 10 mm
1m = 100cm
1km = 1000m

Volume
$1cm^3 = 1000 \text{ mm}^3$
$1m^3 = 1,000,000 \text{ cm}^3$

Area
$1cm^2 = 100mm^2$
$1m^2 = 10,000 \text{ cm}^2$

Distance
1km = 1000m

Example
6 miles = _____ m

> Start with a T chart and place what you are given in the first space, in our case 6 miles.
>
> Place the closest unit conversion you have to where you are going in the next space making sure that what you are getting rid of (miles) is on the bottom and where you are going (km) is on the top.
>
> Repeat step 2 using the next conversion.
>
> **Multiply across the top and the bottom and divide (if necessary).**

6 miles	1.609344 km	1000 m
	1 mile	1 km

$$= \frac{9656.064 \text{ m}}{1} = 9656.064 \text{ m}$$

All conversions can be solved the same way.

Practice 8.1(Round to the nearest hundredth)

1. Convert 5.6 miles to kilometers.

2. Convert 12.2 m to yards.

3. Convert ¾ tons to kilograms.

4. Convert 40°C to degrees Fahrenheit.

5. Convert 10°F to degrees Celsius.

8.2 Area

Area is the size of a surface. Area has units which are squared (for example ft^2 or in^2) Let's start with the formulas for area.

Figure	Area
Rectangle	$A = L \times W$
Square	$A = S^2$
Triangle	$A = \frac{1}{2} bh$
Parallelogram	$A = bh$
Trapezoid	$A = \frac{1}{2} (a+b)h$
Circle	$A = \pi r^2$

Rectangle

The area of a rectangle, $A = LW$

Where, L is the length of the rectangle

And, W is the width of the rectangle.

Width

Length

Example

3cm

12 cm

$A = 12 \times 3$

$A = 36 \text{ cm}^2$

Square

Squares are special cases of rectangles. All the sides of a square are equal.

The area of a square, $A = S^2$

Where, S is the side of the square.

S

Example

Find the area of the square below:

7 in

$A = S^2$

$A = 7^2$

$A = 49\text{in}^2$

Triangle

The area of a triangle, $A = \frac{1}{2}bh$

Where, b is length of the base of the triangle

And, h is the height of the triangle.

Example:

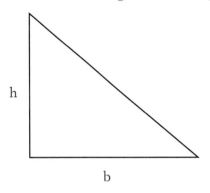

b

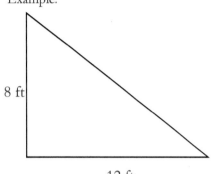

8 ft

12 ft

$A = \frac{1}{2}(12)(8)$

$A = \frac{1}{2}(96)$

$A = 48 \text{ ft}^2$

Parallelogram

The area of a parallelogram, $A = bh$

Where, b is the length of the base of the parallelogram

And, h is the perpendicular height of the parallelogram.

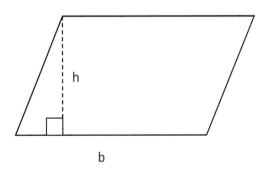

b

Example:

Find the area of the parallelogram.

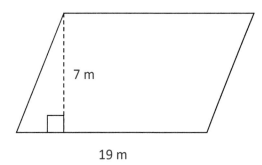

7 m

19 m

$A = bh$

$A = (19)(7)$

$A = 133 \text{ m}^2$

Trapezoid

The area of a trapezoid, $A = \frac{1}{2}(a + b)h$

Where:

- a is the length of one parallel side of the trapezoid
- b is the length of the second parallel side of the trapezoid
- And, h is the perpendicular height (distance between the 2 parallel sides) of the trapezoid

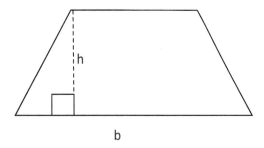

a

h

b

Example

Find the area of the trapezoid:

16 in

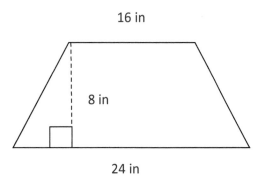

8 in

24 in

$A = \frac{1}{2}(a + b)h$

$$= \frac{1}{2}(16 + 24)8$$
$$= \frac{1}{2}(40)8$$
$$= \frac{1}{2}320$$
$$= 160 \text{ in}^2$$

Circle

The area of a circle, $A = \pi r^2$

Where, r is the radius of the circle

And, π is 3.142 or $\frac{22}{7}$

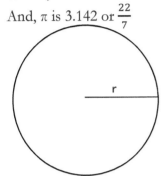

Example

Find the area of the circle below.

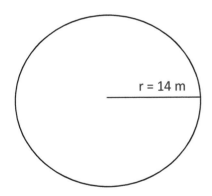

r = 14 m

$A = \pi r^2$

$\quad = \pi 14^2$

$\quad = \frac{22}{7} \times (14)^2$

$\quad = \frac{22}{7} \times 196$

$\quad = 616 \text{ m}^2$

Practice 8.2

1) Find the following:
   - a. Area of square with side 2
   - b. side of square with area 49
   - c. Area of square with diagonal 10
   - d. Area of square with perimeter 36

2) Find the area of the rectangles.
   a.   b.

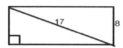 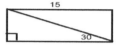

3) Find the area of each triangle.
   a.   b.   c.

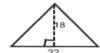 

4) Find the area and perimeter of the following parallelograms.
   a.   b.

5) Given a trapezoid with bases 6 and 15 and height 7, find the median and the area.

6) The bases of a trapezoid are 8 and 22 and the trapezoid's area is 135. Find the height.

7) Find the area of each trapezoid.
   a.   b.   c.

8) Find the areas and circumferences of circles with the following radii:
   a. 1   b. 15

8.3 Perimeter

This section is centered on finding the perimeter (the total distance around) of figures.

Rectangle

The perimeter of a rectangle, P = L + w + L + W

$$= 2(L + W)$$

Where, L is the length of the rectangle

And, w is the width of the rectangle.

Example

Find the perimeter of the rectangle below:

4 in

20 in

P = 2(20 + 4)

= 2(24)

= 48 in

Square

The perimeter of a square, P = L + L + L + L

$$= 4L$$

Where, L is the length of a side of the square.

Example

Find the perimeter of the square below:

15 ft

P = 4(S)

P = 4(15)

P = 60 ft.

Triangle

The perimeter of a triangle, P = the sum of all the sides.

Example

Find the perimeter of the triangle below.

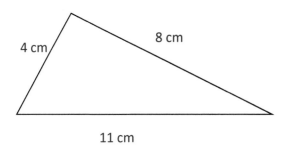

Perimeter (P) = a + b + c

$\qquad$ = 4 + 8 + 11

$\qquad$ = 23 cm

Circle

The perimeter of a circle is called its circumference. The circumference of a circle, C = 2πr or πd

Where, r is the radius of the circle

$\qquad$ d is the diameter of the circle

And, $\qquad$ π is 3.142 or $\frac{22}{7}$.

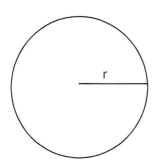
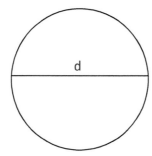

Example

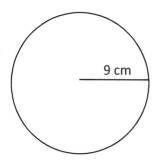

9 cm

Perimeter, P = 2πr

P = 2 x 3.14 x 9

= 57 cm

OR

P = 2 x $\frac{22}{7}$ x 9

= $\frac{396}{7}$

= 57 cm

Practice 8.3

1) How long is each side of a square if the perimeter is 100cm?

2) Find the perimeter of a rectangle that measures 42cm by 19cm

3) Each side of a triangle measures 56cm. What is its perimeter?

4) How wide is a rectangle if the length is 60cm and the perimeter is 180cm?

5) The perimeter of a square is 600cm? How long is each side?

6) I need to buy enough fencing to go all around my rectangular garden. The garden is 30 m long by 4m wide. How many meters of fencing will I need?

7) What is the length of the third side of a triangle if one side measures 10 cm, the second measures 15 cm and the perimeter is 45cm?

Chapter 8 Review

100

1. Find *Area* & *Perimeter* first then cost to *paint* at a rate of $0.23/cm²		2. Find *Area* & *Perimeter* first then the cost to *fence* at a rate of $1269/km	
	4.5 cm		8.5 km
Perimeter	Area	Perimeter	Area
3. Find *Area* & *Perimeter* first then cost to *paint* at a rate of $11.68/m²		4. Find *Area* & *Perimeter* first then the cost to *fence* at a rate of $1310/km	
	5 m		6 km
Perimeter Problem	Area Problem	Perimeter Problem	Area Problem

Determine the area and perimeter of each of the following shapes.

5.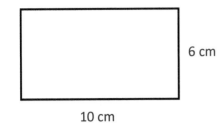

Area =   Perimeter =

6.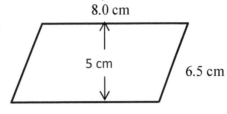

Area =   Perimeter =

7.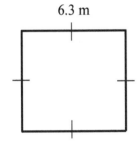

Area =   Perimeter =

8.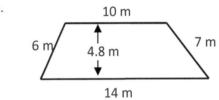

Area =   Perimeter =

9) 1000 mg = _____ g

10) 198g = _____ Kg

11) 8 mm = _____ cm

12) 160 cm = _____ mm

13) 75 mL = _____ L

14) 6.3 cm = _____ mm

15) 109 g = _____ Kg

16) 50 cm = _____ m

17) 5.6 m = _____ cm

18) 250 m = _____ Km

19) 5 L = _____ mL

20) 26,000 cm = _____ m

21) 14 Km = _____ m

22) 16 cm = _____ mm

23) 56,500 mm = _____ Km

24) 1 L = _____ mL

25) 65 g = _____ mg

26) 27.5 mg = _____ g

27) 480 cm = _____ m

28) 2500 m = _____ Km

29) 923 cm = _____ m

30) 27 g = _____ kg

31) 355 mL = _____ L

32) 0.025 Km = _____ cm

(-2)x(+3)

remove 2 groups of +3

For this, you need to make zero pairs

3x2 = 6 (make 6 zero pairs)

take away*

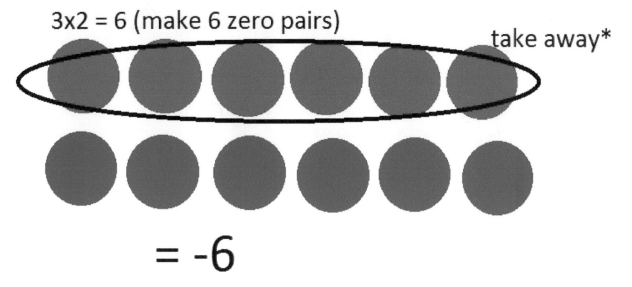

= -6

have 7   owe 3

$(+7) - (-3) = $ (+10)

• = +
○ = −

+7 - -3

+7   -3   = (+10)

+7   -3

0   +10

Chapter 9: Integers

9.1 Integers

Integers can be shown as a set of numbers, as in this example:

$$\{\dots, -3, -2, -1, 0, 1, 2, 3 \dots\}$$

Or integers can be represented on a number line as in this example:

If a number is greater than 0, it is called a positive integer. No sign is needed to indicate a positive integer however you may see a plus sign used at times.

Look at the number line above. All numbers to the right of 0 are positive.

Examples: 4, 8, and 15 are positive numbers all of them appear to the right of 0.

The arrows on the end of the number line indicate that the number line continues to go on forever.

For every positive integer there is an opposite integer which is the negative version of that number.

15 and -15 are opposites

Negative Integers

Negative Integers are any numbers that are less than 0 or and those numbers fall to the left of 0 on the number line. If a number is negative, there will be a negative (-) sign in front of the number.

Example: -12, -8, and -30 are negative integers

One thing to understand about negative numbers is that the farther away from 0 they are, the smaller the number will be.

For Example: -15 is less than -12.

The closer the negative number is to being positive the larger it is!

Absolute Value

Absolute Value is the value of the number without considering the signs. The absolute value of a number is always positive. Absolute value is represented by straight line brackets $|x|$ where x is any number.

Example:

$|17| = 17$

$|-31| = 31$

$|-3| = 3$

If there is a negative sign on the outside of the absolute value sign then that number will be negative once it comes out of the absolute value.

Example:

$-|8| = -8$

9.2 Comparing and Ordering Integers

Negative integers are whole numbers that are less than 0. Since they are less than 0, they always lie to the left of 0 on the number line.

Positive numbers are whole numbers greater than 0 and always lie to the right of 0 on the number line.

Tip: 0 is neither negative nor positive.

If you use this number line to compare numbers, then the numbers increase as you move to the right and they decrease as you move to the left. In other words the highest negative number will come first in order from least to greatest and you will end with the highest positive number.

Example:

Place the following in order from least to greatest

$$-20, -15, 0, 12, 22, -8, 13, -11$$

-20, -15, -11, -8, 0, 12, 13, 22

> Greater Than

< Less Than

≥ Greater Than or Equal To

≤ Less Than or Equal To

These symbols can be used to compare integers.

Example:

Compare the following integers using the symbols <, >, =

1) 19 _____ -12

2) -53 _____ -45

3) -18 _____ -29

Solutions

1) $19 \geq -12$

2) $-53 \leq -45$

3) $-18 \geq -29$

Tips:
- Positive numbers are always greater than negative numbers.
- Negative numbers are always less than positive numbers.
- When using a number line, numbers increase as you move to the right and decrease as you move to the left.

9.3 Adding Integers

There are two rules that you must follow when adding integers. You must look at the signs of each number that you are adding to determine which rule to use.

Same Sign:

When adding signs that are the same you want to add and keep the sign.

For example:

-5 + -6 = -11 all signs are negative

4 + 3 = 7 all signs are positive

Different Sign:

When adding signs that are different, subtract the numbers and keep the sign of the number with the largest absolute value.

Example 1:

-4 + 8 =

8 – 4 = 4

Larger number is 8 and it is positive so the answer is positive.

Example 2:

-6 + 3 =

6 – 3 = 3

The 6 is the number with the largest absolute value and it is positive so the answer (3) will be positive.

Practice 9.3

1) -5 +(-8) = _____

2) 9 + (-14) = _____

3) -16 + 5 =_____

4) 10 + (-4) = _____

5) -12 + (-10) = _____

6) 15 + (-9) = _____

7) -12 + 3 = _____

8) 16 + (-16) =_____

9) 14 + (-7) =_____

10) -8 + (-8) = _____

9.4 Subtracting Integers

There is only one rule that you have to remember when subtracting integers which is to change the subtraction problem to an addition problem.

For subtracting integers only keep the following phrase in mind – Keep, change, change!!!!

Keep the first number exactly the same. Change the subtraction sign to an addition sign. Change the sign of the last number to the opposite sign. If the number was positive change it to negative or if it was negative, change it to positive.

Example:

-6 – 5

= -6 + -5

= -11

15 – (-8)

= 15 + 8

=23

Practice 9.4

1.) 6 - (-7) = _____

2.) -12 - 3 = _____

3.) 10 - (-8) = _____

4.) -9 - 10 = _____

5.) -4 - (-12) = _____

6.) -5 - 9 = _____

7.) 2 - 12 = _____

8.) -13 - (-10) = _____

9.) 6 - (-7) + 8 = _____

10.) 12 + (-15) - (-3) = _____

11.) -8 - 4 - (-7) = _____

12.) 3 - (-7) - 12 = _____

9.5 Multiplying and Dividing Integers

In multiplication and division of integers there is one rule to remember. If the signs are the same then the answer is positive and if the signs are different then the answer is negative.

Examples:

$8(-5) = -40$ {signs are different so the answer is negative}

$-16(-2) = 32$ {signs are the same so the answer is positive}

$45 \div -9 = -5$ {signs are different so the answer is negative}

$-108 \div -9 = 12$ {signs are the same so the answer is positive}

Practice 9.5

1.) $5(-3) =$ _____

5.) $9(-3) =$ _____

2.) $-25 \div -5 =$ _____

6.) $-10(-3) =$ _____

3.) $-8(-3) =$ _____

7.) $81 \div -9 =$ _____

4.) $-42 \div 7 =$ _____

Chapter 9 Review

Solve the following problems.

1. $-2 + (+3) =$

11. $-3(-4) =$

21. $45 - (-27) =$

2. $-5 + (+4) =$

12. $24 \div (-6) =$

22. $19(-4) =$

3. $5 - (-3) =$

13. $5(-18) =$

23. $-42 \div (-6) =$

4. $-7 - (-3) =$

14. $-8 \div (-4) =$

24. $-21 + -19 =$

5. $-14 - 6 =$

15. $17(-4) =$

25. $32 \div (-4) =$

6. $6 + (-8) =$

16. $81 \div (-9) =$

26. $14 - (-7) + (-2) =$

7. $12 + (+7) =$

17. $-21 \div (-7) =$

27. $-8 \cdot -4 \div -2 =$

8. $-8 + (-1) =$

18. $-7(9) =$

28. $-24 \div 4 + -17 =$

9. $-9 - (+6) =$

19. $8(7) =$

29. $7 - (-3) + (-2) - 4 =$

10. $11 + (-2) =$

20. $56 \div (-14) =$

30. $12 + (-7) - (-28) =$

Solve the following word problems using positive and negative numbers.

31. Steve has overdrawn his checking account by $27. His bank charged him $15 for an overdraft fee. Then he quickly deposited $100. What is his current balance?

32. Joe played golf with Sam on a special par 3 course.. They played nine holes. The expected number of strokes on each hole was 3. A birdie is 1 below par. An eagle is 2 below par. A bogie is one above par. A double bogie is 2 above par. On nine holes Frank made par on 1 hole, got 2 birdies, one eagle, four bogies, and one double bogie. How many points above or below par was Franks score?

33. Find the difference in height between the top of a hill 973 feet high and a crack caused by an earthquake 79 feet below sea level.

34. In Detroit the high temperatures in degrees Fahrenheit for five days in January were -12°, -8°, -3°, 6°, -15°. What was the average temperature for these five days?

35. Hightop Roofing was $3765 in the "red"(owed creditors this amount) at the end of June. At the end of December they were $8765 in the "red." Did they make or lose money between June and December? How much?

36. To establish the location of a hole relative to a fixed zero point, a machinist must make the following calculation:

$y = 5 - (3.750 - 0.500) - 2.375$ Find y.

Step 1:　　$k \div 3 = 4$

Step 2:　　$k \div 3 = 4$
　　　　　　　　$\times 3$　　$\times 3$

Step 3:　　$k \div \cancel{3} = \boxed{4 \times 3}$
　　　　　　　　$\cancel{\times 3}$　　　12

Step 4:　　$k = 12$

$6(3x - x) + 2(2x + 3)$

$= (6 \cdot 3x - 6 \cdot x) + (2x \cdot 2 + 3 \cdot 2)$

$= 18x - 6x + 4x + 6$

$= 16x + 6$

Chapter 10:
An Introduction to Algebra

10.1 Real Numbers

The natural or counting numbers are 1, 2, 3, 4,…

The whole numbers consist of the natural numbers and 0: 0, 1, 2, 3, 4,…

The integers consist of the natural numbers together with the negatives and 0: …, -4, -3, -2, -1, 0, 1, 2, 3, 4, …

The rational numbers are numbers that can be written as an integer divided by an integer (or a ratio of integers).

$$\text{Examples:} \quad \tfrac{1}{2} \qquad -\tfrac{1}{4} \qquad 0.19 \qquad 4.27 \qquad 31$$

The irrational numbers are numbers that cannot be written as an integer divided by an integer.

$$\text{Examples:} \quad \sqrt{6} \qquad \pi \qquad \sqrt[4]{9} \qquad e$$

Properties of Real Numbers

Commutative Property

for Addition: $a + b = b + a$

for Multiplication: $ab = ba$

Associative Property

for Addition: $(a + b) + c = a + (b + c)$

for Multiplication: $(ab)c = a(bc)$

Distributive Property

$a(b + c) = ab + ac$ or $(b + c)a = ab + ac$

Additive Identity

$a + 0 = 0 + a = a$

Subtraction is the inverse operation for addition ("undoes" addition) and is the same as adding the negative of the number to be subtracted: $a - b = a + (-b)$

Properties of Negatives

1. $(-1)a = -a$

2. $-(-a) = a$

3. $(-a)b = a(-b) = -(ab)$

4. $(-a)(-b) = ab$

5. $-(a + b) = -a - b$

6. $-(a - b) = b - a$

Multiplicative Identity

$a \cdot 1 = 1 \cdot a = a$

Division is the inverse operation for multiplication ("undoes" multiplication) and is the same as multiplying by the reciprocal: $a \div b = a \cdot \dfrac{1}{b}$

Properties of Fractions

1. $\dfrac{a}{b} \cdot \dfrac{c}{d} = \dfrac{ac}{bd}$

2. $\dfrac{a}{b} \div \dfrac{c}{d} = \dfrac{a}{b} \cdot \dfrac{d}{c}$

3. $\dfrac{a}{c} + \dfrac{b}{c} = \dfrac{a+b}{c}$

4. $\dfrac{a}{b} + \dfrac{c}{d} = \dfrac{ad+bc}{bd}$

5. $\dfrac{ac}{bc} = \dfrac{a}{b}$

6. If $\dfrac{a}{b} = \dfrac{c}{d}$ then $ad = bc$

Practice 10.1

Name the property for each problem below:

1) $x + 9 = 9 + x$

2) $2(x + 3) = 2x + 6$

3) $x + (y + 3) = x + (3 + y)$

4) $(5 y) \cdot (1) = 5 y$

5) $(x y) z = x (y z)$

6) $(x + 5) (7 + x) = (x + 5) \cdot (7) + (x + 5) \cdot (x)$

10.2 Adding Real Numbers

Adding real numbers follows the same rules as adding integers.

To add two numbers with the same sign (both positive or both negative)
- Add their absolute values.
- Keep the sign.

To add two numbers with different signs (one positive and one negative)
- Find the difference of their absolute values.
- Give the sum the same sign as the number with the larger absolute value.

Example
$-32.22 + 124.3.$

$= 124.3 - 32.22$

$= 92.08$

The larger of the two numbers is positive so the answer is positive.

Practice 10.2

1) $-\frac{2}{5} + \frac{1}{2}$

2) $-\frac{2}{5} + \frac{3}{-4}$

5) $\frac{4}{10} + \frac{1}{4}$

6) $-\frac{2}{3} + \frac{3}{4}$

7) $6.3 + -2.2$

8) $-1.7 + -4.3$

3) $\frac{-6}{10} + \frac{2}{3}$

4) $\frac{3}{5} + \frac{1}{-2}$

9) $-9.1 + 5.4$

10) $-0.6 + -2.7$

11) $-7.6 + 1.3$

12) $6.7 + -9.6$

10.3 Subtracting Real Numbers

When subtracting real numbers use the additive inverses or opposites to rewrite subtraction as addition. If adding two numbers with different signs, find the difference between their absolute values and keep the sign of the number with the greater absolute value.

When the greater number is positive, it's easy to see the connection.

$$13 + (-7) = 13 - 7$$
$$= 6 \qquad = 6$$

Example:
-32.3 – (-16.3)

= 48.6

Practice 10.3

1) 3 - 19

2) 7 - (- 4)

3) -5 - 5

4) -23 - 6

5) -2 + (- 4) - 9 + 8 13 14

6) 1 - (- 1) + 6 + (- 6) - 2

7) 6.5 + (- 8.3) - (- 1.6) + 0.7 - 9.9

10.4 Multiplying and Dividing Real Numbers

Zero times any number is zero.	$-1.23\,(0) = 0$
Any number (except zero) divided by zero is undefined.	$\frac{7}{0}$ is undefined
Zero divided by any number (except zero) is zero.	$\frac{0}{7} = 0$
Zero divided by zero is said to be indeterminate.	$\frac{0}{0}$ is indeterminate
The product or quotient of 2 positive numbers is positive.	$7\,(3) = 21;\ \frac{10}{5} = 2$
The product or quotient of 2 negative numbers is also positive.	$-7\,(-3) = 21;\ \frac{-10}{-5} = 2$
The product or quotient of 1 negative and 1 positive number is negative.	$-7\,(3) = -21;\ \frac{10}{-5} = -2$

When writing a negative fraction such as "negative three fourths", all three of these expressions are equivalent: $-\frac{3}{4} = \frac{-3}{4} = \frac{3}{-4}$

Practice 10.4

1) $2.5(-3.5)$

2) $5(6)(-3.2)$

3) $17.4 \div (-3.2)$

4) $\frac{3}{4}\left(-\frac{2}{5}\right)$

5) $\frac{1}{8} \div \frac{-5}{7}$

10.5 Exponents and Order of Operations

The exponent in a number tells how many times to use the number in multiplication or how many times to multiply a number times itself in multiplication.

Example:

$5^3 = 5 \times 5 \times 5 = 125$

Parts of an exponent problem are the actual exponent and the base:

x^2 where x is the base and 2 is the exponent.

Multiplying Exponents:

When multiplying exponents with like bases, add the exponents.

Example:

$x^2 \cdot x^4 = x^{2+4} = x^6$

$5^5 \cdot 5^8 = 5^{5+8} = 5^{13}$

Dividing Exponents:

When dividing exponents with like bases, subtract the exponents.

Example:

$y^5 \div y^3 = y^{5-3} = y^2$

$\dfrac{m^5 n^7}{m^4 n^{10}} = m^{5-4} n^{7-10} = mn^{-3} = \dfrac{m}{n^3}$

Power Rule:

When raising a power by a power multiply the exponents.

$\left(x^4\right)^6 = x^{(4)(6)} = x^{24}$

$\left(3m^2 n^5\right)^4 = 3^4 m^8 n^{20} = 81m^8 n^{20}$

Zero Exponent Rule:

Any base raised to a power of zero is equal to 1.

Example:

$y^0 = 1$

$(4m^8n^2)(-2mn4)^0 = (4m^8n^2)(1) = 4m^8n^2$

Negative Exponent Rule:

It is improper to leave exponents as negative numbers. Take the reciprocal of the number and drop the negative sign.

Example:

$x^{-5} = \dfrac{1}{x^5}$

$3x^{-4} = 3 \cdot \dfrac{1}{x^4} = \dfrac{3}{x^4}$

Order of Operations

When there are multiple operations it is necessary to solve equations in a particular order. Order of Operations (PEMDAS is a special acronym to help you remember)

1. Simplify expressions inside grouping symbols. (parentheses)
2. Simplify or evaluate any terms raised to powers. (exponents)
3. Solve multiplication and division operations in order as you come to them from left to right. (multiply, divide)
4. Finally, do all addition and subtraction operations in order as you come to them from left to right. (add, subtract)

TIP: the words in the parentheses at the end of the lines make "PEMDAS". You can also remember the acronym PEMDAS by remembering the following sentence "**P**lease **E**xcuse **M**y **D**ear **A**unt **S**ally."

Example 1:

Solve the following problem.

$30 - 12 \div 2 + 4 \cdot 3 - 5$

$= 30 - 6 + 4 \cdot 3 - 5$

$= 30 - 6 + 12 - 5$

$= 24 + 12 - 5$

$= 36 - 5$

$= 31$

Since there are no parentheses or exponents you move on to multiplication and division. In this case you will divide first because that is the first option between the two from left to right.

$12 \div 2 = 6$

$4 \cdot 3 = 12$

Now solve the addition and subtraction in order of appearance from left to right.

$30 - 6 + 12 - 5 = 31$

Example 2:

$(8 + 4) \div (10 - 6) + 5^2$

$= 12 \div 4 + 5^2$

$= 12 \div 4 + 25$

$= 3 + 25$

$= 28$

Practice 10.5

1) $15^{-4}(15^8)$

2) $a^7(a^8)(a)$

3) $(3m^4n^6)(2mn)^0(2m^2n)$

4) $-28a^6b^{-3}c^5 \div 7a^{11}b^{-5}c^5$

5) $(-1x^5y^6)^{10}$

6) $\dfrac{(-2ab^7)^3}{(-a^4b^2)^5}$

7) $(5m^3n)(-2mn^3)$

8) $12 \div 4 + 2$

9) $6^2 \div (3 + 9) + 2(5+3)$

10) $\dfrac{6(2+5)}{3(7)}$

11) $26 - [(25 - 11) - 2^3]$

12) $(8^2 - 2^5) \div (24 \div 6) + 3^2$

13) $\dfrac{5(16-5)-1}{4^2-7}$

We begin with the parentheses this time.

$(8 + 4) = 12$ and $(10 - 6) = 4$

Move on to the exponents

$5^2 = 25$

Division

$12 \div 4 = 3$

Addition

$3 + 25 = 28$

10.6 Algebraic Expressions

An algebraic expression is a mathematical expression that consists of variables, numbers and operations. The value of this expression can change.

Steps to Evaluating an Expression:

1. Evaluate an expression by replacing the variable(s) with the given values.
2. Simplify the expression using the order of operations.

Example:

Evaluate the given expression where $x = 8$ and $m = 3$

$4x - 2m$

$= 4(8) - 2(3)$	Substitute in the given values for x and m
$= 32 - 6$	Solve using Order of Operations
$= 26$	

Practice 10.6

Substitute and evaluate: $x = 8, \; y = 6, \; m = 3, \; p = \frac{1}{2}, \; n = \frac{3}{4}$

1. $5y + 8p$

2. $nxy \div m$

3. $2(3x + 6) \div (10m)$

4. $2ny + x$

5. $(x + y) \div p$

6. $6p + 8n$

7. $my - 2x$

10.7 Simplifying Algebraic Expressions Using Properties of Real Numbers

An algebraic expression is one consisting of constants, variables, grouping symbols, and operations. Algebraic expressions are meant to capture the steps of a multi-step calculation without being restricted to fixed numbers all the time.

We simplify algebraic expressions by combining like terms.

- A term is a number or the product of a number and one or more variables raised to a power. Terms are combined using addition or subtraction.
- The number that multiplies the variables in a term is called the coefficient.
- Like terms have the same variable or variables, each raised to the same exponent.
- We can combine like terms into a single term by adding (or subtracting) the coefficients.

2x + 3x can be combined because they have the same variable with the same exponent.

$2x + 3x^2$ cannot be combined because although they have the same variable the variables do not have the same exponent.

Example:

1. Simplify the algebraic expression 2n + 7n by combining like terms.
   When you combine like terms you add the coefficients and keep the variables the same
   2n + 7n = 9n

2. Simplify the algebraic expression $8y^2 - 3 + 2y^2 + 11$ by combining like terms.
   $8y^2 - 3 + 2y^2$

   Our like terms are $8y^2$ and $2y^2$ when combined we get $10y^2$

   $10y^2 - 3$ is our final answer

We can also simplify by distribution using the Distributive Property.

Example: $2(3m + 4) - (8m - 2)$

According to the Order of Operations we do what is in parentheses first however we cannot combine what is in parentheses so we move on to distributing our 2 in through multiplication.

$2(3m + 4) - (8m - 2)$

2(3m) + 2(4) − 8m + 2   Distribute the 2 and the − sign in to each set of parentheses

6m + 8 − 8m +2   Combine like terms

-2m +10

Practice 10.7

Simplify the following.

1) $7x + 5 - 3x$

2) $6w^2 + 11w + 8w^2 - 15w$

3) $6x + 4 + 15 - 7x$

4) $(12x - 5) - (7x - 11)$

5) $(2x^2 - 3x + 7) - (-3x^2 + 4x - 7)$

6) $11a^2b - 12ab^2$

7) $4(7x - 8) + 6(5x + 10)$

8) $6(4x^2 - 5x + 2) + 3(-8x^2 + 11x + 4)$

9) $5(4x^2 - 8x + 3) - 7(6x^2 - 4x + 11)$

10) $4(6x^3 - 4x^2 + 7x + 1) - 9(4x^3 - 2x^2 - 6x + 1)$

11) $10(4x^2 + 8x + 7) - 8(5x^2 + 10x - 9)$

12) $6(4x^2 - 3x + 2) + 5(3x - 6)$

13) $9 (4x^2 - 7x + 12) - 12(3x^2 - 5x - 9)$

14) $4(6x^3 - 4x^2 + 11) - 7(5x^2 + 9)$

15) $3(12x^4 - 16x^3 + 4x^2 - 8x + 24) - 4(9x^4 - 12x^3 - 3x^2 - 6x + 18)$

Chapter 10 Review

Use order of operations to solve the following problems.

1. $18 - (-12 - 3) =$

7. $-19 + (7 + 4)^3 =$

2. $18 + (-7) \cdot (32 - 6) =$

8. $-19 - (-3) + -2(8 + -4) =$

3. $20 + -4(3^2 - 6) =$

9. $-3 + 2(-6 \div 3)^2$

4. $3 \cdot (-4) + (52 + -4 \cdot 2) - (-9.82) =$

10. $2^3 + (-16) \div 4^2 \cdot 5 - (-3) =$

5. $-6(12 - 15) + 2^3 =$

11. $\dfrac{4(-6) + 8 - (-2)}{15 - 7 + 2} =$

6. $-50 \div (-10) + (5 - 3)^4 =$

12. $\dfrac{1.4(4.7 - 4.9) - 12.8 \div (-0.2)}{-4.5 \cdot (-0.53) + (-1)} =$

13) $18 - [5(3 - 1)] + 5$

14) $\dfrac{14 \div 2 + 3 \times 1}{1 + 8 \div 2}$

Simplify

15) $3 - 2(-4) + 3(5)$

16) $(\frac{3}{4})^3$

17) $(-4)^2$

18) $-4^2 - (-3)^2$

19) $-2(.5)^2 + 4(.2)^3$

20) $(-1)^3 - 2(3)^2(-1) - 5$

21) $[5(-2)^2 - 3(4)^2 + 18]^2$

22) $18a - 6b + 2a + 5b$

23) $2x^2 - 3x + 6x - x^2$

24) $2(x - 5) + 3(x + 6)$

25) $3(a - 4) - (2a + 5)$

26) $-2(y - 3) + 6(2y + 5)$

27) $5(x + 2y) - 6(3x - y)$

Identify the property that justifies each statement:

28) $-1(a) = -a$

29) $\frac{1}{2} + (-\frac{1}{2}) = 0$

30) $3(a - b) = 3a - 3b$

31) $x + 5 = 5 + x$

Word Problems → Equations/Ineq.

$$8 + 2x < 20$$
$$-8 \qquad -8$$
$$\frac{2x}{2} < \frac{12}{2}$$
$$\boxed{x < 6}$$
$$-\infty$$

3,4,5

1,2,4

3,2,1

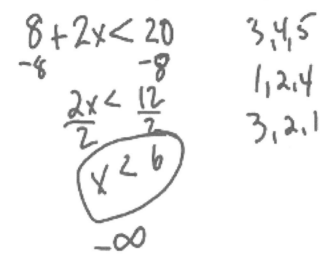

EX3:
$$\left(\sqrt{V-4}\right)^2 = 3^2$$
$$V - 4 = 9$$
$$+4 \qquad +4$$
$$\boxed{V = 13}$$

$$\sqrt{(V-4)(V-4)}$$
$$\sqrt{9 \cdot 9}$$

① Isolate $\sqrt{\ }$
② Square both sides

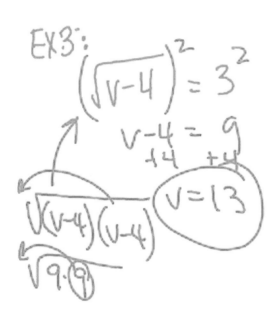

Chapter 11:
Equations and Inequalities

11.1 The Addition Property of Equality

An equation is when two algebraic expressions are set equal to each other.

An algebraic expression represents a multi-step calculation, so setting it equal to something means you know what you want the answer to be, but you don't know what number to put in for your variable to get that answer you want.

Finding that number is called "solving" the equation.

Solving an equation involves "undoing" the calculation represented by the expression to expose the variable.

To "undo" an expression, you must do the opposite of the order of operations. That is to say, you start with adding or subtracting.

The same number may be added to (or subtracted from) each side of an equation without changing the equation.

If $a = b$ Then $a + c = b + c$

Example:

$$\begin{array}{rcl} x - 6 & = & 12 \\ +6 & & +6 \\ \hline x & = & 18 \end{array}$$

In order to solve this equation or get the variable (x) by itself we must move the 6. We can move the 6 by adding a 6 to it because a negative 6 and a positive 6 is 0 so they cancel each other out.

What you do on one side of the equal sign you have to do on the other. So add 12 to 6 and you get $x = 18$

Practice 11.1

1) $y - 5 = -6$

2) $a - 7 = 8$

3) $c - 3 = 9$

4) $x - 14 = 52$

5) $b - 3 = -13$

11.2 The Subtraction Property of Equality

Let a, b, and c represent numbers. Then if a = b Then a - c = b – c.

In other words if you have a problem in which you are trying to solve for the variable and it is an addition problem you would use the Subtraction Property of Equality to do so.

$x + 4 = 6$

$\underline{\quad -4 \quad -4 \quad}$

$x = 2$

> In order to solve this equation or get the variable (x) by itself we must move the 4. We can move the 4 by subtracting a 4 from it because a 4 minus 4 is 0 so they cancel each other out.
>
> What you do on one side of the equal sign you have to do on the other. So subtract 4 from 6 and you get x = 2

Practice 11.2

1) $y + 15 = -6$

2) $a + 7 = 18$

3) $c + 5 = 9$

4) $x + 4 = 52$

5) $b + 8 = -13$

6) $z + -6 = 14$

7) $s + 81 = 101$

8) $x + 32 = -42$

9) $y + 10 = -18$

10) $a + 1 = 4$

11.3 The Multiplication Property of Equality

The Multiplication Property of Equality states that if you multiply both sides of an equation by the same number, the sides remain equal.

$$\text{If } a = b \text{ then } a \cdot c = b \cdot c$$

$$\frac{x}{4} = 3$$

$$(4)\ \frac{x}{4} = 3\ (4)$$

$$\frac{\cancel{4}x}{\cancel{4}} = 12$$

$$x = 12$$

In order to solve this equation or get the variable (x) by itself we must move the 4. We can move the 4 by multiplying both sides by 4. On the left the 4's cancel out

What you do on one side of the equal sign you have to do on the other. So subtract 4 from 6 and you get x = 2

Practice 11.3

1) $\frac{y}{5} = 10$

2) $\frac{x}{7} = 8$

3) $\frac{y}{3} = -4$

4) $\frac{z}{4} = 16$

5) $\frac{y}{-8} = -8$

6) $\frac{a}{-15} = 4$

7) $\frac{y}{8} = 3$

8) $\frac{z}{21} = -2$

9) $\frac{a}{-12} = -4$

10) $\frac{y}{6} = .7$

11.4 The Division Property of Equality

The Division Property of Equality states that if you divide both sides of an equation by the same nonzero number, the sides remain equal.

If a, b and c are any three real numbers such that a = b and c ≠ 0 then the division property of equality is $\dfrac{a}{c} = \dfrac{b}{c}$.

$3a = 6$

$\dfrac{3a}{3} = \dfrac{6}{3}$

$a = 2$

| In order to solve this equation or get the variable (a) by itself we must move the 3. We can move the 3 by dividing both sides by 3. On the left the 3's cancel out.

What you do on one side of the equal sign you have to do on the other. So divide 6 by 3 and you get a = 2. |

Practice 11.4

1) 4x = 12

2) 9x = 9

3) -2x = 14

4) 7x = 21

5) 5y = 35

6) 2y = 13

7) -6x = 45

8) -21x = -63

9) 5a = -50

10) 2a = -100

11.5 Linear Equations

A linear equation is an equation which when graphed produces a straight line. The following are two examples of linear equations with one unknown:

x + 2 = -9

5x + 8 = 6x - 3

When solving linear equations with one unknown follow the steps below:

5x + 8 = 6x - 3

-5x   -5x

    8 = x – 3

   +3   +3

   11 = x

Step 1: Get all of your like terms together with your whole numbers on one side and those with the variable on the other side.

Step 2: Solve for the unknown variable.

Practice:

1. x + 6 = 11

2. 2(3x + 5) = 34 – 2x

3. $\frac{x}{4} + \frac{1}{8} = \frac{5}{8}$

4. 5x + 1 = 31

5. 3x – 1 = 8

6. 7x = 60 + 2x

7. 3x = 72 – 3x

8. 6x + 4 = 20 – 2x

9. 6x + 3 = 23 + x

10. 5x + 4 = 2x + 17

11. 5x + 11 = 20x – 64

12. 28 – x = 17 + 3x

11.6 Formulas

A formula is a special type of equation that shows the relationship between different variables. A formula will have more than one variable.

Examples of Formulas:
x = 2y – 7

$a^2 + b^2 = c^2$

The "subject" of a formula is the single variable that everything else is equal to.

You can change the subject of a formula by rearranging the formula and setting it equal to whatever variable you are looking for.

For Example:

Area (A) = Length (L) x Width (W)

A = LW

Let's solve for width (W) that means get W by itself.

$\dfrac{A}{L} = \dfrac{LW}{L}$

$\dfrac{A}{L} = W$

> In order to get W by itself we must get rid of L
>
> Right now L is being multiplied by W and the opposite of multiplication is division. So in order to get W by itself we would divide by L.
>
> * Remember what you do on one side of the equation you must on the other side.

Example 2:
$s = ut + \frac{1}{2}at^2$, Solve for a

-ut   -ut

$s - ut = \dfrac{1}{2} at^2$

$2(s - ut) = (\dfrac{1}{2} at^2)\, 2$

$\dfrac{2(s - ut)}{t^2} = \dfrac{at^2}{t^2}$

$\dfrac{2(s - ut)}{t^2} = a$

Practice 11.6
1) $C = 2\pi \bullet r$, for r

2) $F = ma$, for a

3) $s = \dfrac{v}{r}$, for r.

4) $d = \dfrac{m}{v}$, for v

5) $I = prt$, for p

6) $A = P + Prt$, for t

7) $s = vt + 16t^2$, for v

8) $A = \dfrac{1}{2}h(a+b)$, for h

9) $S = \dfrac{n}{2}(a+L)$, for n

10) $P = 2(L+W)$, for W

11.7 Problem Solving

Everything that has been studied thus far can be used in problem solving. In this section the focus will be on how to solve different word problems using the S.O.L.V.E. method.

S: Study the problem. This is referred to as "S" the problem. The first step is to highlight the question. The next step is to answer the question "What is the problem asking me to find?" Always start the answer to this question with the word "the."

O: Organize the facts. This step is referred to as "O" the problem. Always "S" the problem first, then "O" the problem. There are three steps to "O."

1) Identify the facts.
2) Eliminate unnecessary facts by crossing them out.
3) List the necessary facts.

L: Line up a plan. "S" and "O" the problem before "L" the problem. In this step choose the operation or operations will be used to solve the problem. Use the facts listed in "O" when writing a plan.

V: Verify the plan with action. Write a numerical expression or equation that can be solved. This is where the actual math work is done and an answer is found.

E: Examine the results. In this step, students will ask the questions:

- "Did you answer what you were asked to find in S?"
- "Is my answer accurate?"
- "Is my answer reasonable?"

Example:

On a shopping trip, Laura and Bo's mother spent the same amount of money as Laura and Bo combined. If Laura spent $37 and their mother spent $64, how much did Bo spend?

S - What is the problem asking me to find? How much Bo spent.

O - Identify and List the Facts: Laura spent $37 and their mother spent $64.

L - Choose the operation(s) necessary to solve: Subtraction

V - Write a numerical equation or expression: $x = 64 - 37$; $x = 27$

E - Did you answer what you were asked to find in S? Yes Bo spent $27

Practice 11.7

1) Three times the greatest of three consecutive even integers exceeds twice the least by 38. Find the integers.

2) The difference of two numbers is 12. Two fifths of the greater number is six more than one third of the lesser number. Find both numbers.

3) In a class containing 58 students, the number of women students is one more than twice the number of men students. How many women students are in the class?

4) An air-conditioner repair bill is $87. This included $30 for parts and an amount for 3 hours of labor. What was the hourly rate that was charged for labor?

11.8 Solving Inequalities

Solving inequalities is not very different from solving equations. In order to solve an inequality use the same steps in solving an equation and isolate the variable.

Example:

Solve $2x + 12 > 20$

First, the variable must be isolated.

$$2x + 12 > 20$$
$$\underline{-12 \qquad -12}$$
$$2x \qquad > 8$$

$$\frac{2x}{2} > \frac{8}{2}$$

$x > 4$

Practice 11.8 Solve the Inequalities

1) $x + 8 \le 12$

2) $\frac{x}{2} > 8$

3) $-2x \ge 8$

4) $-8x \le -64$

5) $-7 < x + 4$

6) $10 > x + 15$

7) $x > 10$

8) $-x \le 25$

Chapter 11 Review

Solve

    1) $-5x - 3 = 12$

    2) $2(2x - 3) = 5x - 11$

    3) $3n - 1 - 4n = 6n - 7 - 4n$

    4) $-(n - 5) + 3(n + 2) = 4(n - 3) - 1$

    5) $6x = 30$

    6) $5x = -40$

    7) $4 + x = 40$

    8) $x + 6 = 17$

    9) $x - 5 = -15$

    10) $-2 = x - 8$

    11) $x / -4 = -6$

    12) $1 + x = x + 1$

    13) $x / 5 = 1$

    14) $w + 14 = -8$

    15) $y + (-10) = 6$

    16) $-11 = a + 8$

    17) $-13 + h = -5$

    18) $-2.3 = x + (-1.1)$

    19) $-7 = -16 - k$

    20) $m - (-13) = 37$

    21) $z + (-13) = -27$

    22) $p - (-27) = 13$

23) $41 = 32 - r$

Set up and write an algebraic equation, then solve:

24) In a class containing 58 students, the number of women students is one more than twice the number of men students. How many women students are in the class?

25) An air-conditioner repair bill is $87. This included $30 for parts and an amount for 3 hours of labor. What was the hourly rate that was charged for labor?

Cumulative Review 6-11

Name of angle	Definition	Example
1) ACUTE		
2) OBTUSE		
3) RIGHT		

Classify each angle:

4) $\angle A = 35°$

5) $\angle B = 100°$

6) $\angle C = 150°$

7) $\angle D = 50°$

8) $\angle F = 90°$

9) $\angle G = 180°$

Use the figure on the right to name each of the following.

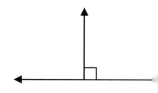

10) Name a pair of complementary angles.

11) Name a pair of supplementary angles.

12) Name a different pair of supplementary angles.

13) Name a linear pair.

Find the measure of each angle

14)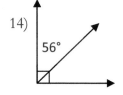

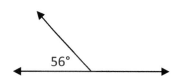

15) Identify the indicated type of triangle in the figure.

a.) isosceles triangles

b.) scalene triangles

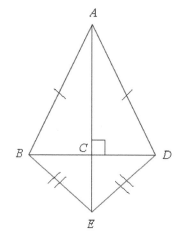

16) Find x and the measure of each side of equilateral triangle RST.

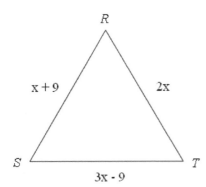

17) Find x, JM, MN, and JN if $\triangle JMN$ is an isosceles triangle

with $\overline{JM} \cong \overline{MN}$.

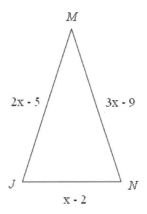

Round to the nearest hundredth:

18) Find the length of the diagonal of a square whose side length is 10 inches.

19) The length of one of the legs in a right triangle is 4 inches. If the hypotenuse is 12 inches long, what is the length of the other leg?

20) Find the length of a diagonal of a square enclosure with a perimeter of 16 feet.

21) Tomato Plants The heights (in inches) of eight tomato plants are 36, 45, 52, 40, 38, 41, 50, and 48.
   a) What is the range of the tomato plant heights?
   b) Find the mean, median, and mode(s) of the tomato plant heights.

22) The populations (in millions) in 2000 on each of the six inhabited continents were 803, 487, 348, 3686, 730, and 31.
   a. What is the range of the populations?
   b. Find the mean, median, and mode(s) of the populations. Round your answers to the nearest tenth.

MASS CONVERSION FACTORS

1gram (g) = 1000 mg   1 kg = 1000 g

LENGTH CONVERSION FACTORS

1 meter (m) = 1000 mm   1 m = 100 cm   1 km = 1000 m

VOLUME CONVERSION FACTORS

1 liter (l) = 1000 ml   1 kl = 1000 l

ENERGY CONVERSION FACTORS

1 calorie (cal) = .001 kcal   1 cal = 4.184 joules

PROBLEMS

23) 5 mg = _____ g

24) 403 g = _____ mg

25) 970 kcal = _____ cal

26) 2.3 ml = _____ l

27) 28.5 mm = _____ m

28) 418.4 joules = _____ cal

29) .560 mg = _____ g

30) 3.5 km = _____ mm

31) For each of the three shapes below, find both the perimeter and area.

	31. Perimeter (m)	32. Area (m²)
8 m 6 m 4 m 10 m		
7 cm 10 cm 18 cm 11 cm	33. Perimeter (cm)	34. Area (cm²)
15 mm 10 mm 12 mm	35. Perimeter (mm)	36. Area (mm²)

	37. Circumference (m)	38. Area (m^2)
25 m		
17 cm	39. Circumference (cm)	40. Area (cm^2)
9 mm	41. Circumference (mm)	42. Area (mm^2)

43) ⁻16 − 8

44) 4 + 22

45) 27 + ⁻42

46) 45 - ⁻35

47) ⁻3 - ⁻30

48) 12 - ⁻42

49) 9. -8 − (-10)

50) 21 – 25

51) -19 – 10

52) 12 – (-1)

53) -25 + x for x = 24

54) x + 20 for x = -14

55) 10 – x for x = -27

56) x – (-3) for x = -34

57) $(-12) \div 4$

58) $(-72) \div 9$

59) $(-30) \div (- (-15))$

60) $(-24) \div (-8)$

61) $35 \div (-7)$

62) $(-225) \div (-5)$

63) $14 \div (-(-7))$

Given $a = 8$, $b = -6$, $d = 3$, $x = -4$, $y = 0.5$.

Evaluate the following algebraic expressions below.

64) $a^2 + 3d^2$

65) $x (y - 2)$

66) $\frac{1}{2} x (y + 0.1)^2$

67) $(- x - d)$

Solve

68) $x + 5 = 12$

69) $p - 10 = 21$

70) $3x + 7 = 11$

71) 10 plus the quotient of a number and 15 is 62

72) Three times a number increased by 6 is 14

73) The sum of three times a number and 7 is 22

Solve

74) $y - 4 = 11$

75) $n + 40 = 25$

76) $3q = 10$

77) $\dfrac{d}{5} = 11$

78) $4x + 5 = 29$

79) $7 - 5x = 22$

80) $8x - 12 + 2x = 72$

81) $6x + x - 12 - 3x = 36$

82) $4x - 3x + 7x - 16 = 24$

83) $4x + 2x + 24 = 36$